BIOCHEMISTRY RESEARCH TRENDS

ACETONE

BIOCHEMISTRY, PRODUCTION AND USES

BIOCHEMISTRY RESEARCH TRENDS

Additional books in this series can be found on Nova's website under the Series tab.

Additional e-books in this series can be found on Nova's website under the eBooks tab.

BIOCHEMISTRY RESEARCH TRENDS

ACETONE

BIOCHEMISTRY, PRODUCTION AND USES

IRENKA KOZŁOWSKA
AND
KLIMEK WALCZAK
EDITORS

nova
science publishers
New York

Additional color graphics may be available in the e-book version of this book.

Library of Congress Cataloging-in-Publication Data

ISBN: 978-1-53613-298-4

Published by Nova Science Publishers, Inc. † New York

CONTENTS

PREFACE

In this collection, the authors develop a mathematical model of kinetics' adsorption in solid microporous mediums. This approach allows for the introduction of two new terms in Fick's law: The first term characterizes the direct adsorption, represented by random function and dependent on the average speed but not on speed itself. The second characterizes the fluctuation of adsorbed quantity and heat dissipation, phenomena which are at the origin of irreversibility adsorption and desorption. Next, residual enzyme activity was investigated to monitor stabilization/destabilization of hen egg-white lysozyme at low, intermediate, and high water content in acetone at 25 oC. The results demonstrate that the stabilization/destabilization of lysozyme depends significantly on water content in acetone. Another study is presented with the goal of providing a thermochemical description of the stabilizing/destabilizing effects of organic solvents on the enzyme stability to elucidate what intermolecular processes produce the main effect on the stability and functions of the enzymes at low, intermediate, and high water content in organic liquids. Following this, the authors examine the part that acetone may play in increasing the use of lignocellulosic biomass through exploitation of all constituents, especially lignin. Examples include the use of acetone in studies of the chemical composition of LCB and the utility of acetone in organosolv pulping and pretreatment of LCB. With the commercialization of biorefineries, it is expected that acetone will continue to play a significant role in the processing of lignin, the most abundant natural aromatic polymer on earth with immense potential to reduce dependence on fossil-based aromatic products. The authors also propose an efficient acetone chemical sensor based on the analytical performances such as sensitivity, limit of detection (LOD), good linearity, and reproducibility. This proposed acetone chemical sensor was introduced by the implementation of ZnO/SnO2/Yb2O3 nanoparticles (NPs) as a successful electron mediator with glassy carbon electrode (GCE) assembly. The concluding paper presents an experimental system of isopropanol-acetone-hydrogen chemical heat pump (IAH CHP) for the recovery of low-grade (<100°C) industrial waste heat. Using coefficient of performance (COP) and exergy efficiency as evaluation indices, the system performance is systematically evaluated based on the experimental data.

Chapter 1 - The gas adsorption phenomenon in a microporous solid is certainly a transport phenomenon. Each adsorbed molecule brings with it an energy quantity owing to its mass centre movement and inter atomic vibrations and rotations. The molecule leaves a large part of this energy to the adsorbent solid once it is fixed. The adsorption of a gas by a microporous solid is an exothermic phenomenon. The released heat will have an effect on the adsorption rate. In addition, the overheating phenomenon may be local, which implies a fluctuation of the adsorption rate in the superheated portion of the volume, and not on the total volume of the adsorbent solid. We also know that there are three different resistances to the transfer of heat and mass. These observations show that: it is impossible to describe the interactions between the released heat and the adsorption rate by a single equation without the introduction of random functions. The developed theories that are similar to the chemical kinetics or that are from Knudson diffusion need a serial development in the differential equation by the introduction of several diffusivity constants which will be neglected in the following because of the model complexity. The Fick's diffusion theory is the most suitable for such phenomena despite its drawbacks such as: Constant adsorbent volumic-mass, constant diffusivity D, the fluid drainage pipe not being curly. Such drawbacks are repaired in our model by the introduction of the fluctuation which is observed experimentally and predicted by Fick's theory and by assuming that each molecule is adsorbed directly before undergoing the diffusion and the fluctuation effects. In this work, we develop a mathematical model of the kinetics' adsorption in solid microporous mediums. This approach allows us to introduce two new terms in Fick's law: The first term characterizes the direct adsorption, represented by random function. It depends on the average speed but not on speed itself. The second characterizes the fluctuation of adsorbed quantity and heat dissipation, phenomena which are at the origin of the irreversibility adsorption and desorption. In our model, we have two kinds of adsorption, the surface adsorption (macroporous and lateral surface) and the volume filling in the microporous solid. Moreover, we make a distinction between the diffusion and transport phenomena; the particles are initially transported in the macroporous solid then they are diffused in the microporous. To understand the mechanism of gas adsorption by microporous solids, physical kinetics is the required theory. It allows to determine all the parameters that have an effect on the adsorption rate. We have determined: the minimal distance from which the interaction forces adsorbate-adsorbent orients the adsorbate molecule to the adsorbent; the existence of a critical temperature and its effect on the relaxation time relative to the adsorbed matter quantity; the underscoring of adsorbed matter fluctuation in which the interval is active. The importance of the model and all the possibilities allowing a deeper study of the mechanism of gas adsorption by a microporous solid are well shown in the following.

Chapter 2 - Numerous benefits of employing enzymes in non-aqueous organic liquids exist, including catalysis of industrially-important synthetic reactions, suppression of undesirable side reactions caused by water, and high solubility of hydrophobic reagents. Residual enzyme activity was investigated to monitor stabilization/destabilization of hen egg-white lysozyme at low, intermediate, and high water content in acetone at 25°C. The advantages of our methods are as follows:

1. Residual enzyme activity curves can be determined in the entire range of water content in organic liquids.
2. Enzyme activity values can be measured at fixed reaction conditions.
3. Information about changes in the state of the catalytically-active sites can be obtained separately from other contributions (e.g., changes in the solvation of reagents and products in water – organic mixtures).

Our results clearly demonstrate that the stabilization/destabilization of lysozyme depends significantly on water content in acetone.

1. At high water content, residual activity values are close to 100%.
2. A minimum on the residual activity curve was observed at intermediate water content. Acetone augments the irreversible inactivation of lysozyme in this region.
3. Lysozyme demonstrates significant residual activity in water-poor acetone.

Chapter 3 - The aim of our investigation is to provide a thermochemical description of the stabilizing/destabilizing effects of organic solvents on the enzyme stability to elucidate what intermolecular processes produce the main effect on the stability and functions of the enzymes at low, intermediate, and high water content in organic liquids. Bovine pancreatic α-chymotrypsin was utilized as a model enzyme. Isothermal calorimetry measurements were performed to study the stability of α-chymotrypsin in water-acetone mixtures. Interaction enthalpies of α-chymotrypsin with water-acetone mixtures have been measured at 25^0C. Our results clearly show that isothermal calorimetry is an effective experimental tool for studying the simultaneous action of water and organic liquids on the stability of enzyme macromolecules. Thermochemical results were compared with the enzyme activity data. The degree of stabilization/destabilization of α-chymotrypsin depends significantly on the water content in acetone. There are three concentration regimes. At high water content, the residual enzyme activity values are close to 100%. At intermediate water content, residual enzyme activity is close to zero. The residual catalytic activity in the water-poor acetone is ~80%.

Chapter 4 - In recent times, lignocellulosic biomass (LCB) was mainly used to produce cellulosic pulp for paper and other cellulose products, such as cellulose derivatives. The pulping industry relies on inorganic agents, such as sodium hydroxide and sodium sulfide, that are combined to make the "white liquor" utilized in the kraft process, the dominant chemical pulping process used today. Organic solvents are rarely used for pulping, even though organosolv processes have numerous advantages over kraft pulping, including better selectivity resulting in higher yields and less negative environmental impact. One important example of an organosolv process, the Alcell process, is based on ethanol, but acetone has been suggested for use as well. The Alcell process was implemented in a pilot plant facility based on hardwoods for about a decade at the end of XXth century producing attractive byproducts including lignin. However, organosolv pulping has not been broadly used for various reasons, including the inferior strength properties of organosolv pulp when compared to those of kraft pulp and negative results of economic analysis based on comparisons with larger scale kraft pulping operations. With increasing concerns over the depletion of fossil-based resources and the adverse effects that the combustion of fossil fuels may have on the global climate, there is a renewed interest in lignocellulosic biomass (LCB) and studies are underway to utilize all of the LC constituents, in addition to cellulose, to produce energy, chemicals, and materials in second generation biorefineries. Whereas starch- and sugar-based biomass such as corn and sugar cane are feedstocks for first generation biorefineries; lignocellulosic biomass such as wood agriculture residues and grasses provide the feedstocks for second generation biorefineries. In this context, the processes that were used in the past should be re-examined and improved and new ones should be developed (for example the process of acetone-butanol-ethanol production by fermentation of sugars recovered by hydrolysis of lignocellulosic polysaccharides). The inorganic agents that are commonly used in conventional pulping processes may also be used in second generation biorefineries to decrease the recalcitrance of LCB in alkaline or acid based pretreatment processes, but organic solvents, especially renewable solvents such as ethanol and acetone, are expected to play a more prominent role in second generation biorefineries than in traditional pulping processes. The part that acetone may play in increasing the use of lignocellulosic biomass through exploitation of all constituents, especially lignin, is examined here. Examples will include the use of acetone in studies of the chemical composition of LCB and the utility of acetone in organosolv pulping and pretreatment of LCB. Acetone is routinely used in solvent extraction procedures because its solubility parameter is compatible with that of lignin, making it an excellent choice for use in studies of the fractionation and purification of lignin as well as the production of different sizes of lignin, including nanoparticles, for higher value applications of lignin. With the commercialization of biorefineries it is expected that acetone will continue playing an important role in the processing of lignin, the most abundant natural aromatic polymer on earth with immense potential to reduce dependence on fossil-based aromatic products.

Chapter 5 - Acetone (2-propanone or dimethyl ketone) is the simplest and most important of the aliphatic (fat-derived) ketones, which generally used as organic solvent of industrial and chemical significance. Pure acetone is a colorless, somewhat aromatic, flammable, mobile liquid that boils at 56.2°C (133°F). Construction of highly efficient toxic chemical sensors is the key approach for the determination of carcinogenic chemicals in the environment and ecosystem. We report here, an efficient acetone chemical sensor based on the analytical performances such as sensitivity, limit of detection (LOD), good linearity, and reproducibility. The proposed acetone-detecting electrode was introduced by the implementation of ternary metal oxides nanoparticles (TMO NPs) as a successful electron mediator with glassy carbon electrode (GCE) assembly. The prepared TMO NPs were well crystalline-doped nanomaterial (NMs) and produced by implementation of hydrothermal procedure at low temperature. The conventional methods such as Fourier-transform infrared spectroscopy (FTIR), ultraviolet visible spectroscopy (UV-vis), field emission scanning electron microscopy (FESEM), X-ray photoelectron spectroscopy (XPS), energy-dispersive X-ray spectroscopy (EDS), and powder x-ray diffraction (XRD) were utilized for characterization of prepared NPs in terms of morphological, elemental, optical and structural properties. The linear dynamic range (LDR) of 0.34 nmolL^{-1} to 3.4 mmolL^{-1} with lower LOD (S/N = 3) and a higher sensitivity were exhibited by lab-made fabricated sensor based on TMO NPs for selective acetone detection. In shortly, the TMO NPs are utilized as an excellent electron mediator with Nafion/GCE assembly in a chemical sensor for acetone detection even at the very low concentration. Therefore, the chemical sensor is fabricated with TMO NPs may be a promising highly sensitive sensor by reliable I-V detection method for the effective detection of hazardous and carcinogenic chemicals in medical as well as health-care fields.

Chapter 6 - This paper presents an experimental system of isopropanol-acetone-hydrogen chemical heat pump (IAH CHP) for the recovery of low-grade (<100°C) industrial waste heat. Using coefficient of performance (COP) and exergy efficiency as evaluation indices, the system performance is systematically evaluated based on the experimental data. The performance of the IAH CHP is found to be greatly affected by the endothermic and exothermic temperatures, and the flow rate of working fluid. In addition, the economic performance of the IAH CHP is analyzed as well. The results indicate that the long service-life of IAH CHP can significantly improve the system's economic performance, compensating for the high initial investment. The local energy price is found to have impact on the CHP's economic performance. If the local price of electricity is sufficiently high, the CHP may have better economic performance than other comparable energy systems.

Chapter 7 - Acetone has a longstanding use in organic chemistry labs as the solvent of choice for the cleaning of glassware. This application extends to biological labs, where acetone may be used to rinse away organic contaminants, or water following water-based cleaning. Acetone waste from glassware cleaning can be readily recycled by fractional distillation. Its low boiling point (56°C) permits its effective separation from water and other high boiling substances, while small amounts of other volatile organics do not usually interfere with its repeated use in cleaning. Acetone has generally not been the solvent of choice for other applications in organic chemistry including chromatography on silica gel, where ethyl acetate in combination with hexane or other saturated hydrocarbon(s) is typically used as the mobile phase. Despite its limited use, acetone has certain advantages over the commonly used ethyl acetate in chromatography. First, it is typically less expensive. While acetone is generally considered to be UV transparent only above about 300 nm, it actually has a window of low absorbance in the 205 to 220 nm range. In contrast, ethyl acetate absorbs strongly up to about 250 nm. Acetone as solvent along with a non-absorbing saturated hydrocarbon co-solvent permits UV detection of eluent from chromatography at or near 210 nm, a much shorter wavelength that possible with ethyl acetate. This allows UV detection of a wider range of eluents, including simple alkenes. Acetone is generally considered to have a slightly greater eluting power than ethyl acetate in chromatography on silica gel. Acetone is an especially attractive alternative to ethyl acetate for chromatography if one wishes to recover and separate the solvent mixture by distillation, due to its lower boiling point (56°C) vs 77°C for ethyl acetate. Use of acetone in conjunction with heptane is especially practical, as the two solvents can be separated sufficiently for repeated use by fractional distillation using a standard automated commercial distillation instrument. This application of acetone has significant potential for reducing solvent consumption and waste generation in organic chemistry labs.

In: Acetone
Editors: Irenka Kozłowska et al.
ISBN: 978-1-53613-298-4

Chapter 1

Studies of the Kinetics Adsorption of Acetone in Microporous Solids

Foued Mhiri*
Department of Physics, Monastir University, Monastir, Tunisia

Abstract

The gas adsorption phenomenon in a microporous solid is certainly a transport phenomenon. Each adsorbed molecule brings with it an energy quantity owing to its mass centre movement and inter atomic vibrations and rotations. The molecule leaves a large part of this energy to the adsorbent solid once it is fixed. The adsorption of a gas by a microporous solid is an exothermic phenomenon. The released heat will have an effect on the adsorption rate. In addition, the overheating phenomenon may be local, which implies a fluctuation of the adsorption rate in the superheated portion of the volume, and not on the total volume of the adsorbent solid. We also know that there are three different resistances to the transfer of heat and mass. These observations show that: it is impossible to describe the interactions between the released heat and the adsorption rate by a single equation without the introduction of random functions.

* Corresponding Author Email: ribat005@Yahoo.fr.

The developed theories that are similar to the chemical kinetics or that are from Knudson diffusion need a serial development in the differential equation by the introduction of several diffusivity constants which will be neglected in the following because of the model complexity. The Fick's diffusion theory is the most suitable for such phenomena despite its drawbacks such as: Constant adsorbent volumic-mass, constant diffusivity D, the fluid drainage pipe not being curly.

Such drawbacks are repaired in our model by the introduction of the fluctuation which is observed experimentally and predicted by Fick's theory and by assuming that each molecule is adsorbed directly before undergoing the diffusion and the fluctuation effects.

In this work, we develop a mathematical model of the kinetics' adsorption in solid microporous mediums. This approach allows us to introduce two new terms in Fick's law: The first term characterizes the direct adsorption, represented by random function. It depends on the average speed but not on speed itself. The second characterizes the fluctuation of adsorbed quantity and heat dissipation, phenomena which are at the origin of the irreversibility adsorption and desorption.

In our model, we have two kinds of adsorption, the surface adsorption (macroporous and lateral surface) and the volume filling in the microporous solid. Moreover, we make a distinction between the diffusion and transport phenomena; the particles are initially transported in the macroporous solid then they are diffused in the microporous.

To understand the mechanism of gas adsorption by microporous solids, physical kinetics is the required theory. It allows to determine all the parameters that have an effect on the adsorption rate. We have determined: the minimal distance from which the interaction forces adsorbate-adsorbent orients the adsorbate molecule to the adsorbent; the existence of a critical temperature and its effect on the relaxation time relative to the adsorbed matter quantity; the underscoring of adsorbed matter fluctuation in which the interval is active.

The importance of the model and all the possibilities allowing a deeper study of the mechanism of gas adsorption by a microporous solid are well shown in the following.

Keywords: adsorption, diffusion, fluctuation

NOMENCLATURE

A accumulation due to direct adsorption in the mean and the large pores (mol $kg^{-1}s^{-1}$);

$\frac{1}{b}$	relaxation time relative to pressure in isobar conditions (s);
D	diffusivity constant of micro pores ($m^2 s^{-1}$);
F	proportionality constant;
i	Complex number $i^2 = -1$;
$n(r_c, t)$	Number of adsorbed molecules in ($mol\ kg^{-1}$) by spherical volume of r_c radius at time t;
$n(r,t)$	average density of particles in point at time t ($mol\ kg^{-1}$); $n(r_c, \infty) = n_\infty = \lim_{t\to\infty} n(r_c, t)$ at 25 °C
P	pressure (Pa), assumed to be constant;
r	instantaneous particle position in spherical coordinates (m);
r_c	radius of spherical grain of adsorbent (m);
t	time (s);
T	temperature (°K);
V	volume (m^3)
$\frac{1}{\omega^2}$	relaxation time relative to temperature (s);
$\frac{1}{\omega_c^2}$	relaxation time relative to temperature T_c (s);

Greek Letters

Δ	Laplatian operator;
α	integrate constant (m^{-1}), $\frac{1}{\alpha}$ minimal distance from interaction adsorbat-adsorbent;
δ	integrate constant (without unity),
τ	Relaxation time relative to fluctuation – energy dissipation(s)

1. Introduction

The development of the industrial applications of the adsorption phenomenon requires a good knowledge of the heat and mass transfers in the adsorbent mediums. Consequently, the determination of the parameters which control these transfers and especially the kinetics of adsorption are interesting in several fields: Chemistry, Biology, Pharmaco-chemistry, environmental applications and molecules separation (molecular sieves).

The new industrial applications of machines using the adsorption phenomena in their functioning need the development of mass and heat transfer theory. The kinetic adsorption is one of the primordial theories for the development of these machines. Many models relative to the kinetic adsorption exist in the literature [1, 2]. These models differ with respect to their hypothesis or their experimental conditions. Some assume adsorption to be a diffusion phenomenon in homogenous and isotropic media and use Fick's law [2, 3]. Others have highlighted the effect of forces, of the Van Der Walls type, on the adsorption speed [4-7]. They have shown that the molar interaction energy rises with the adsorbate mass contained in the medium, and that the increase of the adsorbate-cation interactions is in agreement with the cations concentration in the alkaline series.

These deductions introduce the heterogeneity notion of interaction energy in the system. Moreover, the thermo-kinetic experiments confirm the existence of spontaneous fluctuation processes [8-15].

We propose a new mathematical formulation based on the mass balance of the adsorbed matter and the interaction effects between the adsorbed quantity and energy in adsorption. We take into account the heterogeneity [16], anisotropy [17] and spontaneous fluctuation in kinetic adsorption. An analytical expression has been obtained and compared to the measurements realized on the couple: Activated carbon-acetone.

Many equations have been proposed but their application is still limited to some concentration intervals [18]. Other models ignore some system aspects or contradict some already known results. Some other equations lack consistency and engage some experimental parameters instead of defining them dynamically, which bereaves the process of dynamic comprehension [19]. In the proposed models, the number of constants is generally greater than three [20-24].

In general, the three phenomena: adsorption, desorption and diffusion act at the same time [25]. To describe diffusion, some authors use Knudson diffusion theory [26] which requires the definition of many diffusivity constants. Since the number of diffusivity is large, namely the gas diffusivity, surface diffusivity, surface intrinsic diffusivity, theoretical self-diffusivity and apparent diffusivity. This huge number shows the problem complexity as well as the number of variables to be defined.

The gas transportation in a microporous solid is controlled by an active diffusion and a molecular sieve mechanism. It is mainly influenced by the sorption features, the adsorbat molecule geometry and the micropore structure [26, 27]. It is known that the adsorption-desorption phenomena aren't reversible and adsorption is exothermic, the heat should be dissipated by thermal transfer inducing thermal resistances in heat transfer, which will necessarily have an effect on the adsorption kinetics.

Because of the problem complexity, it is preferable to use Fick's diffusion [28] and model through a matter balance, which allows to summarise the variables on which the adsorption depends.

In this work, we propose a new base of adsorption kinetics of gas in a microporous solid. We start by the basic hypothesis: existence of the diffusion (Fick's law); direct adsorption (effect of the surface nature) and existence of the partial adsorption (fluctuation phenomenon). We establish the differential equation from these basic hypotheses. Then, we solve it analytically in isobar and isotherm conditions. The solution contains only three constants to be determined experimentally, and in non-isotherm conditions. Then, we compare our model values with the experimental ones, which confirms the model for different pairs (activated carbon-ethanol and activated carbon-acetone). Moreover, the defined links between the dependent constants are verified experimentally, which constitutes a new validation of our theoretical model. The proposed model is complicated and its resolution was built in a spherical coordinate system [24, 26] which fits the spherical shape of microporous grains. By a variable separation, the differential equation relative to spatial coordinates is hypergeometric (non usual equation) [29]. Once the mathematical burden is overcome, the application becomes relatively easy as the solution is a series among which the number of terms to be used is chosen depending on the computing admissible error.

2. Theoretical Study

2.1. Determination of the Equation

The gas adsorption in a microporous solid is a transportation phenomenon which is a process that witnesses the existence of transfer or global transportation of matter, energy or movement quantity in macroscopic scales [28]. The main physical aspects of these phenomena can be described by similar methods and are characterised by a first order time propagation equation which is Fick's law [28] in simple cases.

$$\frac{\partial n}{\partial t} = D \Delta n \tag{1}$$

where D is a constant that depends on the situation and n is the field corresponding to a particular transportation phenomenon. Generally, many phenomena are described by equations more complicated than (1), which is considered a first approximation.

Diffusion takes place while decreasing the concentration directions, and therefore tends to make uniform the molecular distribution of the diffusing substance in all the allowed space. Therefore, this tendency must be considered in a statistical or microscopic way as there may be local fluctuations which occur for short time intervals: the molecular flow in certain areas can be reversed [28]. In general, the transportation phenomena are described by a more complex equation. To be convinced, one has to check the transportation equation with production and adsorption the diffusion with a chemical reaction (molecule producing and vanishing) where the balance in a nuclear thermal reactor is written as: Accumulation = gain per diffusion + production – adsorption; or mathematically:

$$\frac{\partial n}{\partial t} = D\Delta n + P - X \tag{2}$$

To solve this equation, we should know in each case the natures of the production P and the adsorption X.

The industrial microporous solids are composed of micro, medium and large pores. The closest representation of these solids is the following (Figure 1):

The adsorption in these solids, from the adsorption potentials, is of two kinds:

- The adsorption in the micropores (Volume filling);
- The adsorption in medium and large pores.

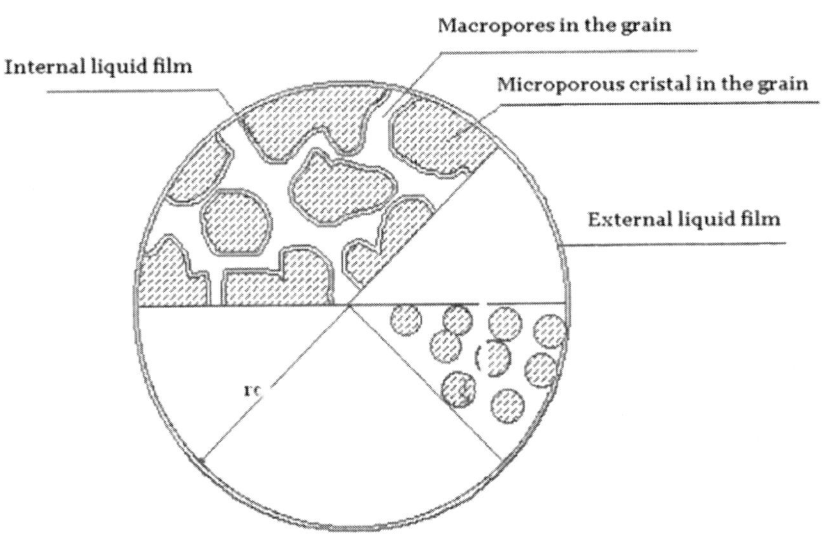

Figure 1. Spherical representation of a microporous solid.

The gas adsorption in a microporous solid is due to three microscopic phenomena: diffusion; adsorption and fluctuation due to an instantaneous local desorption.

From Figure 2, we note that there is a direct adsorption flux in the mean and large pores and on the external surface of the adsorbent solid grain represented in index 1. Index 2 represents the flux of diffused matter inside the micropores and index 3 represents the back flux of flux 2 because of elastic impacts between solid edges and adsorbed molecules. Moreover, it contains the diffused flux in opposite directions to 2 by spherical symmetry. The superposition of flux 2 and 3 creates an instantaneous fluctuation of adsorbed quantity in the contact volume.

From the model in Figure 2, we can see that the differential equation of the system internal representation is the result of a balance of the adsorbed matter unit mass.

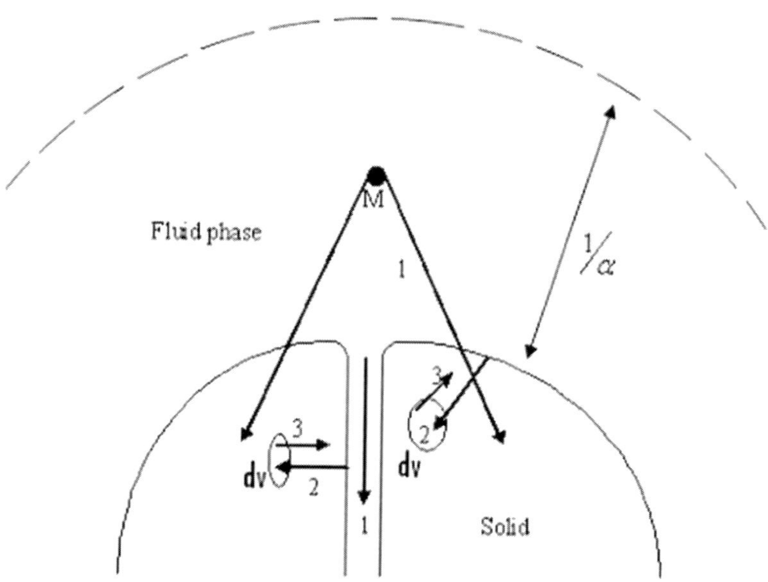

Figure 2. Physical model with dv is an overheated elemental volume1: direct adsorption (transportation phenomenon); 2: diffusion; 3: fluctuation.

Accumulation adsorbate = gain by diffusing in the adsorbent interior + direct adsorption in the mean and large pores and on the external surface + matter exchange due to fluctuation (Figure 2). The differential equation can be written as:

$$\frac{\partial n(r,t)}{\partial t} = D\,\Delta n(r,t) + A(r,t) + B(r,t) \tag{3}$$

In the differential equation $\frac{\partial n(r,t)}{\partial t} = D\,\Delta n(r,t) + A(r,t) + B(r,t)$, the functions n and B depend on the spatial coordinates, pressure, temperature and time. By integration on the system volume, we obtain:

$$\frac{\partial \oiiint_V n\,dv}{\partial t} = D\oiiint_V \Delta n\,dv + \oiiint_V A\,dv + \oiiint_V B\,dv.$$

The quantity $\oiiint_V n\,dv$ =n (Volume, P,T,t) doesn't depend any more on coordinates but on the volume and the ratio $\frac{n(\text{volume}, P, T, t)}{n_\infty(\text{same volume}, P, T, t)}$ is independent of the volume. To solve this equation, one has to know in each case the nature of direct adsorption A(r,t) and that of fluctuation B(r,t).

The fluctuation phenomenon has been considered, for a long time, odd till the middle of the 20th century thanks to the development of the out of equilibrium thermodynamics [8, 9] based on the fundamental theorem of Onsager [10, 11] which has become the basis to understand the spontaneous appearance of the order in the out of order. Particularly, the application of non-equilibrium thermodynamics leads to an oscillating process [15].

To model the term corresponding to fluctuation B(r,t), we bring in the dissipation fluctuation theorem [19, 20]. This latter expresses the fact that if we observe the dissipation in a system, we have to observe fluctuation.

Similarly, the fluctuation observation for the collective variables of a complex system implies the existence of dissipation. In fact, dissipation is roughly an irreversible energy transfer from macroscopic or collective freedom degrees to internal freedom degrees which are enormous.

This theorem confirms the irreversibility between adsorption and desorption and the effect of resistors to thermal transfers on the kinetics adsorption. Moreover, it enables us to choose the best model for the fluctuation term which will be written as: B(r,t) = Z(t) R(r) by a variable separation. Function Z(t) characterises the system fluctuations during time; it expresses the deep link between dissipation and fluctuation in the system. As a first approximation and from [19, 20], for a first approximation, function Z(t) can be considered constant (Z(t) = C) in isotherm and isobar conditions and fluctuation is written as B(r,t) = C R(r).

2.2. Determination of Direct Adsorption

From the model represented in Figure 1, each molecule has to be adsorbed directly in the adsorbent solid, and then, it undergoes the effects of diffusion and fluctuation which act on adsorption speed. According to [28] accumulation, term A can be written as:

$$A = b\,n = V_{moy}\,\Sigma_a\,n$$

Term b, which is assumed constant in isobar and isotherm conditions, depends on the adsorption macroscopic efficient surface and means speed of molecules in the gas around the microporous solid. If we accept that this gas is perfect, we prove that pressure is given by:

$$p = \frac{1}{3} N m V_{moy} \text{ where } V_{moy} = \left(\frac{3}{Nm}\right)^{1/2} p^{1/2}$$

and $b = V_{moy}\,\Sigma_a = \left(\frac{3}{Nm}\right)^{1/2} p^{1/2}\,\Sigma_a$

The direct adsorption term is given by A(r,t) = b n(r,t). Equation (1) is rewritten as:

$$\frac{\partial\, n(r,t)}{\partial t} = D\,\Delta n(r,t) + b\,n(r,t) + C\,R(r) \tag{4}$$

The adsorbent is composed of spherical grains with a mean radius r_c of a micrometer size. The system has a spherical symmetry and only the spatial variable appears in the differential equation. By a variable separation, we seek a solution whose form is n(r,t) = Y(t) R(r)with initial and limit conditions: $n(r,0) = n_0$; $n(r_c, \infty) = n_\infty$ and $\left(\frac{\partial n}{\partial r}\right)_{r=0} = 0$.

2.3. Resolution of the Differential Equation

The study of adsorption potentials with respect to the pore radium is given by Figure 3:

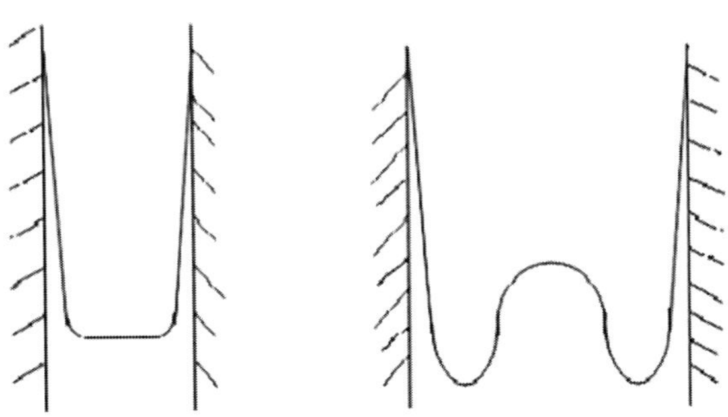

(a) (b)

Figure 3 (a). Variation of adsorption potential in micropores, (b) Variation of adsorption potential in medium and large pores.

The study of adsorption potentials shows that the adsorption is done in the following order:

- Adsorption in the micropores;
- Adsorption in the medium pores;
- Adsorption in the large pores.

The desorption follows the reverse process.

From Figure 3 representation, the distribution of pores relative to their diameters is random. The filling is also done randomly, which depends on the distribution of pores in the grain with respect to their diameters. Therefore, there is not any link between the position of the particle in the microporous solid and time t. Otherwise, there is no real concentration front in the microporous solid, which enables to use the variable separation method to solve the differential equation.

$$\frac{\partial\, n(r,t)}{\partial t} = D\,\Delta n(r,t) + b\,n(r,t) + C\,R(r)$$

We seek the general solution in the form n(r,t) = Y(t) R(r), one obtains the system of differential equations according to:

$$\frac{dY(t)}{dt} + \omega^2\, Y(t) = Z(t) \tag{5}$$

$$\frac{d^2R(r)}{dr^2} + \frac{2}{r}\frac{dR(r)}{dr} + \alpha^2\, R(r) = 0;\, r \in [0; r_c] \tag{6}$$

with ω^2 is a positive real and $\alpha = \left(\frac{b+\omega^2}{D}\right)^{1/2}$

The solution of equation (5) is written [30] as:

$$Y(t) = Y(0)\, e^{-\omega^2 t} + e^{-\omega^2 t} \int_0^t e^{\omega^2 \tau}\, Z(\tau)\, d\tau \qquad (7)$$

$$\frac{d^2R}{dr^2} + \frac{2}{r}\frac{dR}{dr} + \left(\frac{b+\omega^2}{D}\right) R = 0$$

The resolution method is given by [29]:
$\frac{d^2R(r)}{dr^2} + \frac{2}{r}\frac{dR(r)}{dr} + \alpha^2 R(r) = 0$ with α^2 is a strictly positive real.

Carrying out the variable change: $R(r) = r\, e^{-\frac{r}{2}}\, u(r)$

$$\frac{dR}{dr} = e^{-\frac{r}{2}} [(1 - \frac{r}{2})\, u + r \frac{du}{dr}]$$

$$\frac{d^2R}{dr^2} = e^{-\frac{r}{2}} [r \frac{d^2u}{dr^2} + (2-r)\frac{du}{dr} + (\frac{r}{4} - 1)\, u]$$

The equation (6) becomes:

$$r \frac{d^2u}{dr^2} + (2-r)\frac{du}{dr} + (\frac{r}{4} - 1)\, u + (\frac{2}{r} - 1)\, u + 2\frac{du}{dr} + \alpha^2\, r\, u = 0$$

$$\frac{d^2u}{dr^2} + \frac{4-r}{r}\frac{du}{dr} + \frac{r^2(1+4\alpha^2) - 8r + 8}{4r^2}\, u = 0 \qquad (8)$$

By analogy with $\frac{d^2u}{dr^2} + \frac{\tilde{\tau}(r)}{\sigma(r)}\frac{du}{dr} + \frac{\tilde{\sigma}(r)}{\sigma^2(r)}\, u = 0$ (9)

Which is a generalized equation of the hypergeometric type in which $\sigma(r)$ and $\tilde{\sigma}(r)$ are polynomials of a non-superior degree to 2, and $\tilde{\tau}(r)$ a polynomial of a non-superior degree to 1. We meet equations of this type by resolving Laplace and Helmholtz equations in curvilinear coordinates by separation of the variables, in the fundamental problems of the quantic mechanics: a particle movement in a spherical symmetry field, harmonic oscillator, search for equations solutions of Schrôdinger, Dirac and Klein-Goudon for the Colombian potential, particle movement in an electric or homogenous magnetic field [29]. Equation (9) appears equally in a good number of model problems of the atomic, molecular and nuclear physics.

Equations of the type (9) admit, as particular solutions, special functions belonging to the following classes:

- orthogonal classical polynomials (polynomials of Jacob, Laguerre and Hermite);
- spherical functions;
- cylindrical functions;
- hypergeometrical functions.

These functions are often called special functions of the mathematical physics.

The final result is written as:

$$n(r_c,t) = 4\pi\, Y(t) \sum_{j=1}^{\infty} 2^j \frac{(i\alpha)^j}{(j+1)!} e^{-i\alpha r_c} (j!)(j+1) \sum_{\gamma=0}^{j} \frac{(-2i\alpha)^{j-\gamma-1}}{\gamma![(j-\gamma)!]^2 (j-\gamma+1)} [r_c^{j-\gamma+2} +$$

$$\frac{j-\gamma+2}{i\alpha} r_c^{j-\gamma+1} + \frac{(j-\gamma+2)(j-\gamma+1)}{(i\alpha)^2} r_c^{j-\gamma} + \dots\dots\dots + \frac{(j-\gamma+2)!}{(i\alpha)^{j-\gamma+1}} r_c + \frac{2(j-\gamma+2)!}{(i\alpha)^{j-\gamma+2}} \tag{10}$$

A first order development (j = 1) in the case of Z(t) constant gives:

$$\frac{n(r_c,t)}{n_\infty} = Y(t) \times$$

$$\left[\left(r_c^2 + \frac{1}{\alpha^2} \right) \cos(\alpha\, r_c) + \left(\frac{\alpha}{2} r_c^3 - \frac{2}{\alpha} r_c \right) \sin(\alpha\, r_c) \right] +$$

$$\mathrm{i}\,Y(t)\times\left[\left(\frac{\alpha}{2}r_c^3-\frac{2}{\alpha}r_c\right)\cos(\alpha\, r_c)-\left(r_c^2+\frac{1}{\alpha^2}\right)\sin(\alpha\, r_c)\right] \tag{11}$$

with $Y(t) = e^{-\omega^2 t}\,[\,Y(0) - \dfrac{C}{\omega^2}\,] + \dfrac{C}{\omega^2}$ (12)

For $Z(t) = C$ and with limit and initial conditions

$$Y(t) = \frac{C}{\omega^2}\left[1 - \frac{n_\infty - n_0}{n_\infty}e^{-\omega^2 t}\right] \tag{13}$$

If adsorbent is initially dry $n_0 = 0$

$$Y(t) = \frac{C}{\omega^2}\left[1 - e^{-\omega^2 t}\right] \tag{14}$$

This expression has been already considered by other authors in the form $1 - H\, e^{-\frac{t}{\lambda}}$ where λ is the relaxation time and H is a constant [25].

$$\frac{n(r_c,t)}{n_\infty} = \delta\times\left(1-e^{-\omega^2 t}\right)\times$$
$$\left[\left(r_c^2+\frac{1}{\alpha^2}\right)\cos(\alpha\, r_c)+\left(\frac{\alpha}{2}r_c^3-\frac{2}{\alpha}r_c\right)\sin(\alpha\, r_c)\right]+$$
$$\mathrm{i}\,\delta\times\left(1-e^{-\omega^2 t}\right)\times\left[\left(\frac{\alpha}{2}r_c^3-\frac{2}{\alpha}r_c\right)\cos(\alpha\, r_c)-\left(r_c^2+\frac{1}{\alpha^2}\right)\sin(\alpha\, r_c)\right] \tag{15}$$

where Y(t) is given by the differential equation [30]:

$$\frac{dY}{dt} + \omega^2 Y = Z(t) \tag{16}$$

with ω^2 a positive constant in the isotherm case.

Equation (7) is a first order linear differential equation with constant coefficients whose solution is:

If z(t) not constant [30]:

$$Y(t) = Y(t_0)\, e^{-\omega^2 (t-t_0)} + e^{-\omega^2 (t-t_0)} \int_{t_0}^{t} Z(\theta)\, e^{-\omega^2 (t_0-\theta)}\, d\theta \tag{17}$$

Function Z(t) can be a general form and without any additional hypothesis, it is difficult to precise the properties of the system described by such an application. Therefore, we have to assume some hypothesis for which the representation may have more flexible mathematical forms. Equation (9) is a physically-acceptable solution of (8) if Y(t) has to be finite for each t and has to converge when t heads to infinity.

With equation (17), we add the following hypothesis: the interactions of energy fluctuation-dissipation; matter fluctuation-energy fluctuation and energy dissipation-matter fluctuation vary linearly.

For $t_0 = 0$ equation (17) becomes:

$$Y(t) = Y(0)\, e^{-\omega^2 t} + e^{-\omega^2 t} \int_{0}^{t} Z(\theta)\; e^{\omega^2 \theta}\, d\theta \tag{18}$$

This satisfies the causality.

2.3.1. Correlation Z(t)

To have the solution given by (18), one has to determine Z(t). In this paper Z(t) will be considered as the dependence factor between fluctuation and dissipation. To determine this factor, we have to answer the following questions:

- How do we determine correctly the sense of the studied dependence?

- Does a dependence exist between the studied variables and how do we measure the degree of such a link?
- How do we obtain the correct form of the dependence Z on n or t which enables, in the given case, to quantify the influence of Z on n and formulate the imposed conclusions? [31].

The study by applying a statistical probabilistic one described here has confirmed the hypothesis of an influence of the fluctuations on the adsorption speed. In the isotherm case, Z is a function of n and n is a function of t so that Z = Z(t). Function Z defines the deep link between fluctuation of matter and energy dissipation. This implies that n(t) is a random variable as many non-controllable phenomena may affect its values. Z is a random variable too and since Z is a function of n, the problem consists in determining the correlation between Z and time t.

The deduction of the corresponding empirical equation which links both variables can be split into two steps:

- Establishment of the general form of the sought functional relation;
- Computing of the parameter estimates of the equation which fit as much as possible the experience data.

It's worth noting that the first step is less familiar than the second as no resolution methods are available in the literature [31].

The most used and studied dependences are the linear regression and parabolic regression. In our case, these regressions don't provide the wished precision degree. So, we have to try other dependence forms namely the hyperbolic, exponential, or of power forms. In the case of the study of the adsorption kinetics and for the dependences between the fluctuation and the dissipation, these dependences can be formulated as linear ones via the variable change. The dependence which describes our experiments is the exponential one.

2.3.1.1. Study of Function Z(t)

This function expresses the deep link between fluctuation and energy, and matter dissipation, it isn't measurable because of its random behaviour. Moreover, there is no mathematical study which enables to determine its expression. This function is mathematically-inaccessible. Only some properties can be described from the experiment. In the isotherm case, adsorption Z(t) satisfies the following properties:

- It depends on the adsorbed molecule number n(rc,t), on the volume available at time instant t and on the temperature.
- It is finite for any quantity of the adsorbed matter.
- Its first derivative is finite for any adsorbed matter quantity.
- When the system goes to saturation, this function tends to be constant. There's no volume available for the matter fluctuation, and the heat diffusion permanent regime is established.
- Initially, when the microporous solid is empty of adsorbat, this function must be null. If there's no adsorbat in the adsorbent, there's no fluctuation.

Function Z = Z(n) and n = n(rc,t) are functions of time. We can express Z as a function of time Z = Z(t). Moreover, our study handles the kinetics and it is irrelevant to express all functions with respect to the same variables. Mathematically, in the isotherm case these properties are:

- The limit of Z(t) exists for any value of t.
- The limit of the first derivative exists for any value of t.
- It goes to a constant value once t goes to infinity and its first derivative goes to zero.
- Initially (t = 0) this function has to be null and its first derivative should be finite.

From these mathematical properties, the form Z = Z(t) in the isotherm case is given by Figure 4.

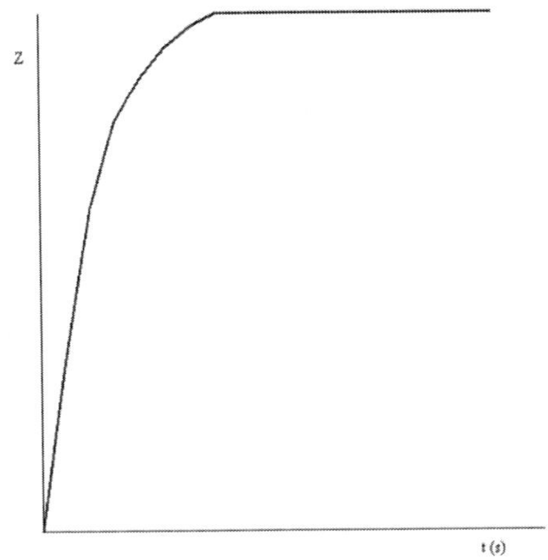

Figure 4. Z = Z(t).

The most effective interaction on the adsorption speed is the following: the interaction between energy dissipation and matter fluctuation and the interaction between matter fluctuation and energy fluctuation. In such a case, function Z(t) can be written as: Z(t) = h(t) + g(t) where h(t) is the function that characterises the link between energy dissipation and matter fluctuation and g(t) characterises the link between matter fluctuation and energy fluctuation.

2.3.1.2. Study of h(t)

At the start of adsorption (t small), the available volume for the adsorbed a molecule is big, which enables an easy movement of molecules: matter fluctuation is important. However, at the end of adsorption (t big), the adsorbent solid is close to saturation and the volume available for the free moving of molecules is so small that matter fluctuation is near zero, which enables us to write h(t) as: $h(t) = E\,(1 - e^{-\frac{t}{\tau}})$ where $(1 - e^{-\frac{t}{\tau}})$ characterises the mean values of matter fluctuation with respect to time and E is a constant (in isotherm case) which characterises the mean dissipation of energy.

2.3.1.3. Study of g(t)

In the beginning of adsorption, energy fluctuation has to be important because of the high number of hits between the molecules of adsorbat and the adsorbed molecules and microporous solid edges since the available volume is relatively big and the molecules have a high degree of movement freedom. Once the solid micropores are filled, the available volume for the movement of molecules decreases as well as their free movement degree, which leads to a reduction of the energy fluctuation. Moreover, the heat conduction by the microporous solid tends to the steady state, which enables to characterise energy fluctuation by the function $e^{-\frac{t}{\tau}}$. In the isotherm case, the mean value of the fluctuation of matter due to energy fluctuation is $F(1-e^{-\omega^2 t})$ where $(1-e^{-\omega^2 t})$ the adsorbed quantity characterises during the time and F the percentage of molecules whose positions are fluctuated because of energy fluctuation. Function Z(t) is written as:

$$Z(t) = F e^{-\frac{t}{\tau}} (1-e^{-\omega^2 t}) + E(1-e^{-\frac{t}{\tau}}) \tag{19}$$

By hypothesis, the interactions between the different fluctuations are linear and from equation (18), function Y(t) is written as the product of two functions as:

$$Y(t) = C(1-e^{-\frac{t}{\tau}})(1-e^{-\omega^2 t}) \tag{20}$$

The term $(1-e^{-\omega^2 t})$ describing the links between matter and energy dissipations and fluctuation and from previous reasoning has an effect only on small concentrations. This result can be proved theoretically. We prove that this term is different from unity if $\frac{n(t,T)}{n_\infty(T)} \langle \frac{1}{2}$. The product of the two exponential terms in the expression of Y(t) is familiar in adsorption kinetic theories [19, 35].

2.4. Existence of Fluctuation

In statistical thermodynamics, entropy is defined as a value which measures the degree of a system disorder in a microscopic level. If Ω is the number of microstates, the cell number of the corresponding space at a thermodynamical state; the Boltzman postulate, to which Plank has given its final form, establishes the link between the entropy and Ω as $S = f(\Omega)$ where f is a universal function. The unique solution of this equation is $f(\Omega) = K \ln\Omega$, so that:

$$S = K \ln\Omega$$

In the case of the adsorption of gas in a microporous solid, Ω is the number of possibilities to arrange the adsorbed molecules over the total number of the receiver sites $n_\infty(T)$:

$$\Omega = \frac{n_\infty(T)!}{n(t,T)!(n_\infty(T) - n(t,T))!}$$

with Stirling formula, we have:

$$S = K\left[n_\infty(T)\ln(n_\infty(T)) - n(t,T)\ln(n(t,T)) - (n_\infty(T) - n(t,T))\ln(n_\infty(T) - n(t,T))\right]$$

The entropy is:

$$\frac{\partial(\frac{S}{K})}{\partial n} = 0 \Rightarrow -\ln(n(t,T)) + \ln(n_\infty(T) - n(t,T)) = 0 \Rightarrow n(t,T) = \frac{n_\infty(T)}{2}$$

Function S = f (n) has a maximum for $n(t,T) = \frac{n_\infty(T)}{2}$, entropy S decreases as we have a negative temperature.

From this study the term $(1-e^{-\frac{t}{\tau}})$ which links fluctuations to dissipations has an effect once $\frac{n(t,T)}{n_\infty(T)} \langle \frac{1}{2}$. If this ratio equals $\frac{1}{2}$, the term $(1-e^{-\frac{t}{\tau}})$ is set to unity. By hypothesis, the interactions between the different fluctuations are linear and from equation (5), function Y(t) is written as the product of two functions as:

$$Y(t) = \delta\,(1-e^{-\frac{t}{\tau}})(1-e^{-\omega^2 t}) \text{ with } Y(0) = 0. \qquad (21)$$

2.5. Existence of a Characteristic Temperature

The couple is characterised by an internal parameter which is its resistance to matter transfer, this resistance may burn out at T_c and its value changes. Otherwise, resistance to matter transfer can have two values: one for $T < T_c$ and the other for $T > T_c$. The causality differential equation which satisfies the variation curve of ω^2 with respect to $\frac{T}{T_c}$ is:

$$\frac{d\omega^2}{\omega^2} = f\,\frac{dT}{T} \qquad (22)$$

and its integration yields:

$$\omega^2 = \omega_c^2 \left(\frac{T}{T_c}\right)^f \qquad (23)$$

where ω_c^2 and f are two constants that can be determined experimentally. This theory has been confirmed by the papers [32, 33, 34].

3. EXPERIMENTAL STUDY

The measurements are collected at the National High School of industrial techniques and Mines of Nantes (France) within the Department of Energetic and Environment Engineering.

3.1. Experimental Condition

The measurements of the adsorbed mass are carried out in a fluidised bed in isotherm conditions (Figure 5). The used adsorbent is the activated carbon PIC AN C60, the physicochemical characteristics are presented in Table 1.

Table 1. Characterization of activated carbon NC 60

Origin	Coconut	Oxidation Temperature	203 °C
Activation	CO_2+H_2O (à 900°C)	Volumic mass	1.4 $g.cm^{-3}$
Specific surface	1240 m^2g-^{1}	Specific heat	0.8-0.9 $J.g^{-1}.K^{-1}$
Porous volume	0.55 cmg^{-1}	/	/

The grain radius is between 0.3 and 0.6 mm; the bed mass is 0.3 kg; the bed height is 0.12 m and the reactor diameter is 100 mm. All the tests are carried out with a solvent concentration at the input of 30 g m^{-3}. All the tests were performed at a U_0 speed = 2 U_{mf}. The speed minimal of fluidization measured according to the temperature U_{mf} is given in Table 2.

Table 2. The speed minimal of fluidization measured according to the temperature U_{mf}

T = 30 °C	T = 50 °C	T = 75 °C	T = 100 °C
0.23	0.210	0.195	0.18

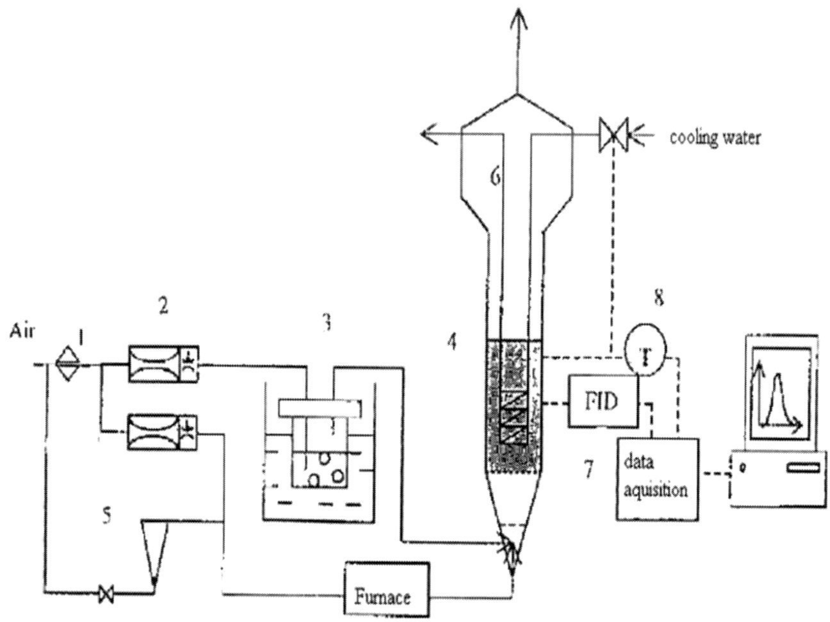

Figure 5. Representative diagram of the installation utilized for the experimental tracing of isotherms. 1: filter; 2: Mass volumetric governor; 3: Wash-bottle; 4: fluidized bed reactor; 5: Conical float rotameter; 6: Exchanger with cool water; 7: Detector with ionization of flame; 8: Thermocouple K.

The active coal particles are washed with distilled water, then placed in a drying oven at 105 °C during 24 good fortunes before fluidizing them to wash them and remove the fine impurities.

The reactor has been dimensioned by taking into account the physical properties and the minimal speed of the particle fluidisation of the activated carbon. The tubular oven with 0.81w power allows to heat the gas at the input for an isotherm functioning. An air temperature control is achieved in a wind box of the column by a PID generator (Eurotherm

2132) acting on the uphill power. The cold water exchanger enables to evacuate the yielded energy during adsorption. It's composed of 18 copper spirals and a PID regulator which allows to regulate water flow depending on temperature in the activated carbon bed. The steam saturated air flow is generated by the pinching of a dry air flux in a liquid solvent. The pinching system is thermostat to avoid the concentration fluctuation of the saturation of ambient temperature. This flux saturated in the solvent is diluted in a wind box by a pure air current.

By considering the variation of the partial molar flow of the adsorbate between the column input and output equivalent to the adsorbed adsorbate quantity per unit of time, we can write the following relation:

3.1.1. Input – Output = Accumulation

$$G_0\,(X_0 - X) = M_{bed}\,\frac{dn}{dt}$$

with $X = \dfrac{x}{1-x}$ and $x = \dfrac{C_0}{M}\dfrac{RT}{P}$

We deduce n

$$n = \frac{X_0 G_0}{M_{bed}} \int_{t=0}^{t=t_f} \left(1 - \frac{X}{X_0}\right) dt$$

By integrating the previous equation, we obtain the value of n.

3.2. Application of the Theoretical Model

The theoretical curves are plotted according to equation (15), with Z(t) equal to a constant and n(r,t) = 0, Figure 6.

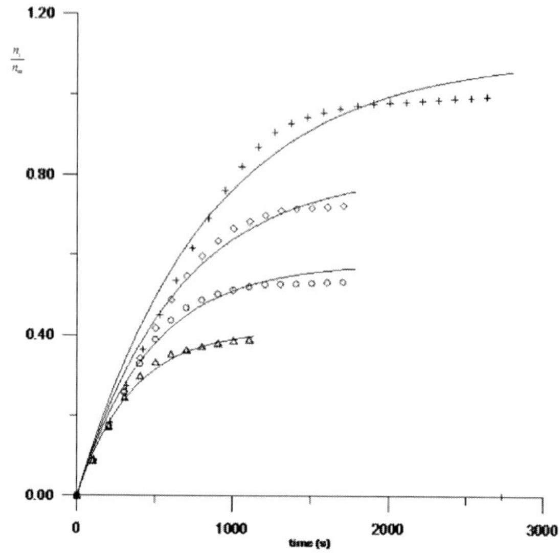

Figure 6. Evolution of quantity of acetone adsorbed in activated-carbon(—) theoretical curves, + experimental points; + 30 °C, ◊: 50, 75 and Δ at 100.

Table 3. Experimental values of the constants

T °C	30	50	75	100
δ **of activated carbon-acetone**	9.90	9.56	9.36	8.93

Table 4. Experimental values of constants

Pair	T_c (°C)	f	ω_c^2 (s^{-1})	b (s^{-1})	D (m^2s^{-1})
Activated carbon -acetone	66	4.013	$1.900\ 10^{-3}$	$-1.928\ 10^{-5}$	$3.214\ 10^{-2}$

The kinetics of acetone adsorption in the active carbon has been analysed numerically by our model and compared to experimental data. It comes out that the determined model fits the reality despite the isobar adsorption assumption. Moreover, we can affirm that gas transportation in microporous solids is controlled by an active diffusion and molecular sieve mechanism. The latter depends essentially on the adsorbat molecule geometry, on the pore structure and the adsorption characteristics.

4. Comparative Study between Different Models

4.1. Analytical Models

The understanding of the adsorption phenomenon necessarily involves kinetics. So, it is essential to master the parameters of the kinetics. The proper choice of the adsorption model that correctly describes the adsorption properties of the couple used (adsorbate/adsorbent) is essential.

The main kinetic models we have used to treat the phenomenon successfully. To explain it:

- Model 1: pseudo first order model;
- Model 2: pseudo second-order model;
- Model 3: Fick's Law model;
- Model 4: Model based on the improvement of Fick's Law.

4.1.1. Pseudo First Order Model

The pseudo first order assumes that the rate of the occupied adsorption sites is proportional to the number of the unoccupied sites. The pseudo first order equation was expressed by [36].

$$q_t = q_e(1 - e^{-k_1 t}) \qquad (24)$$

with k_1: adsorption rate constant for the first order (s^{-1})

4.1.2. Second Order Pseudo Model

McKay and Ho presented a model to characterize the adsorption kinetics by taking into account both the case of rapid fixation adsorbate at the most reactive sites and that of slow fixation at the low sites energy. It is given by the following equation [36].

$$q_t = \frac{q_e^2 k_2 t}{1 + q_e k_2 t} \tag{25}$$

with k_2: the adsorption rate constant for the pseudo second order (g mg^{-1} s^{-1}).

4.1.3. Model of Fick's Law

Fick's scattering model is used to estimate the kinetic adsorptions of adsorbent/adsorbate pair. Considering a particle has a spherical shape, the adsorption rate can be estimated using the diffusion equation:

$$\frac{\partial n}{\partial t} = \frac{1}{r^2}\frac{\partial}{\partial r}\left(r^2 D \frac{\partial n}{\partial r}\right) \tag{26}$$

The empirical solution of equation (26) is given by the following equation [37]:

$$\frac{n(r_c, t)}{n_\infty} = 1 - \frac{6}{\pi^2}\sum_{n=1}^{\infty}\frac{1}{n^2}\exp\left(-\frac{n^2\pi^2 Dt}{r_c^2}\right) \tag{27}$$

The expression given in equation (27) converges quickly after a long time or approaching saturation.

Equation (27) can, then, be simplified [37]:

4.1.4. Model Based on the Improvement of Fick's Law

The equation is written in this form [32, 33, 34]: A development of order one (j = 1) for Z (t) is non-constant, and n (r, 0) = 0.

$$\frac{n(r_c,t)}{n_\infty} = Y(t)\times\left[\left(r_c^{\,2}+\frac{1}{\alpha^2}\right)\cos(\alpha r_c)+\left(\frac{\alpha}{2}r_c^{\,3}-\frac{2}{\alpha}r_c\right)\sin(\alpha r_c\right]+$$
$$iY(t)\left[\left(\frac{\alpha}{2}r_c^{\,3}-\frac{2}{\alpha}r_c\right)\cos(\alpha r_c)-\left(r_c^{\,2}+\frac{1}{\alpha^2}\right)\sin(\alpha r_c)\right]$$

$$\mathrm{Y(t)} = \delta\,(1-\mathrm{e}^{-\frac{t}{\tau}})(1-\mathrm{e}^{-\omega^2 t})$$

α, ω^2, τ are characteristic of the adsorption phenomenon of gas by microporous solids. Only the real part of equation (7) is considered:

$$\frac{n(r_c,t)}{n_\infty} = \delta\left(1-e^{-t/\tau}\right)\left(1-e^{-\omega^2 t}\right)\left[\left(r_c^{\,2}+\frac{1}{\alpha^2}\right)\cos(\alpha r_c)+\left(\frac{\alpha}{2}r_c^{\,3}-\frac{2}{\alpha}r_c^{\,2}\right)\sin(\alpha r_c)\right] \tag{28}$$

4.2. Adjustment

The developed theoretical study allowed us to find a mathematical expression that expresses the number of molecules adsorbed in a spherical volume of radius r_c as a function of time, knowing that the mean radius is r_c = 0.45mm. The objective of the theoretical elaboration is to give a physical sense to the different terms that the model contains, and thus, to facilitate the understanding of the adsorption process at the molecular level. This theoretical basis will enable us to better interpret and deduce information on the physical adsorption of an adsorbate/adsorbent couple. The model is retained as an adequate model, based on the best R^2 coefficient of fit. This Coefficient is obtained by numerical simulation of the experimental isotherms while using the Microcal ORIGIN version 8 software. This coefficient, by its value, measures the quality of the adjustment which takes values between 0 and 1, obtained after numerical simulation of the experimental isotherms. We have, therefore, been able to determine the various characteristic parameters for describing and interpreting, at the molecular scale, the adsorption-adsorbate adsorption process. The best model is the one for which R^2 is very close to 1; it means that the theoretical model reflects the experimental points [38].

We modeled the four adsorption isotherms (acetone) using these kinetic models. In the tables below, we present the average values of the R^2 adjustment coefficients of each model.

4.3. Adjustment Result for Each Model

4.3.1. Adjustment Curves of the Pseudo First Order Model

In this section, we show the kinetic isothermal adjustment curves for the different couples studied at each temperature, as well as the values of the error relative to each pair. Numeric values are grouped in tables.

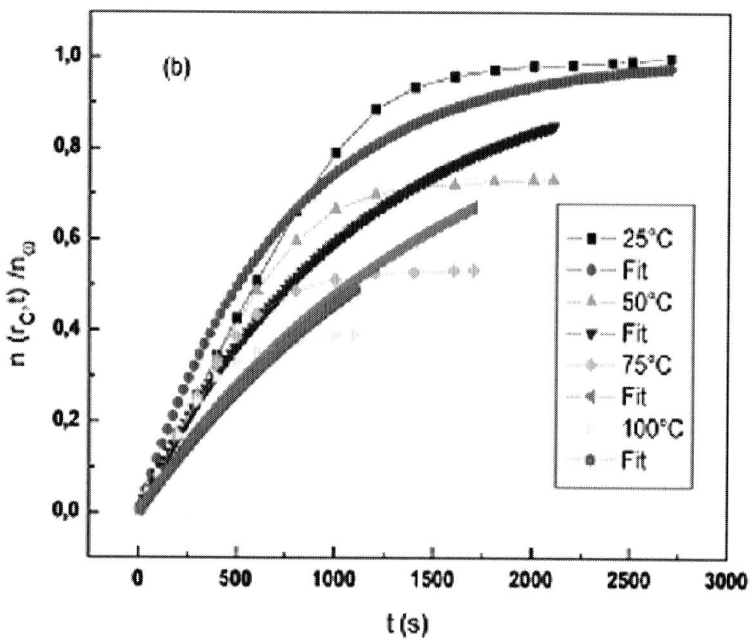

Figure 7. Pseudo first order model fitting curves: Acetone-activated carbon.

Table 5. Values of the adjustment coefficients R^2 for each kinetic model (adsorbent: activated carbon)

Pair	Temperature(°C)	30	50	75	100
Acetone-activated carbon	Model 1	0.978	0.954	0.831	0.825
	Model 2	0.871	0.977	0.936	0.901
	Model 3	0.775	0.652	0.788	0.699
	Model 4	0.996	0.996	0.999	0.998

4.3.2. Parameter Adjustment of Pseudo First Order Model

Table 6. Parameter adjustment of pseudo first order model

Pair	Temperature (°C)	$k_1(s^{-1})$
Activated carbon-acetone	30	0.00315
	50	0.0020
	75	0.0017
	100	0.0014

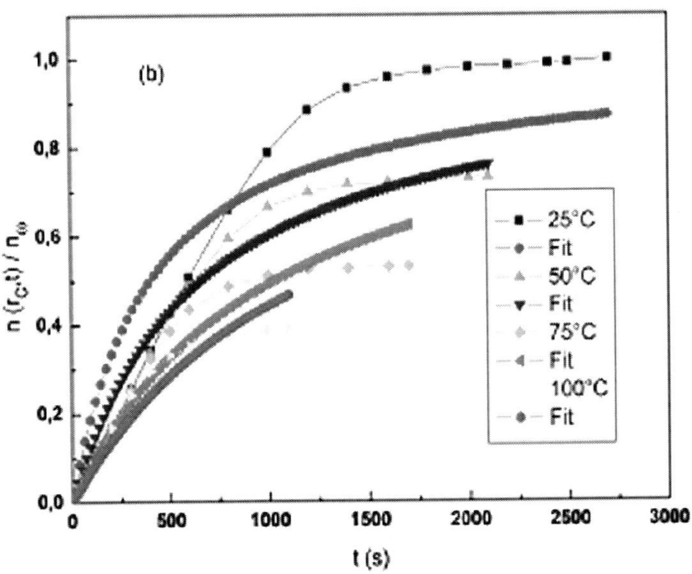

Figure 8. Pseudo-second order model adjustment curves: acetone.

4.3.3. The Relative Errors of the Pseudo First Order Model

Table 7. The relative errors of the pseudo first order model

T (°C)	time (s)	$(n_r/n_\infty)_{exp}$	$(n_r / n_\infty)_{th}$	$\lvert [(n_r/n_\infty)_{exp} - (n_r/n_\infty)]/(n_r/ n_\infty)_{exp}\rvert$
30	10	0,00385	0,01	0.6
	1063	0.69509	0,5128	0,2
	1809	0,95237	0,8828	0,5
50	10	0,00435	0,00895	0.6
	1005	0,66514	0,59431	0,19
	2000	0,73007	0,83393	0,14
75	10	0,00385	0,00334	0,13
	1001	0,35434	0,28553	0,19
	2000	0,38696	0,48783	0,2
100	10	0,00436	0,00608	0,3
	500	0,33112	0,26516	0,19
	1000	0,38996	0,45671	0,17

4.3.4. Adjustment Curves of the Pseudo Second-Order Model

This section presents the pseudo-second-order model fit curves for the different pairs as well as the error values for each pair, and the adjustment results are grouped in the following tables.

Table 8. Parameter adjustment of pseudo second order model

Pair	Temperature (°C)	k_2 (g mg s^{-1})
Activated carbon-acetone	30	0.0022
	50	0.0021
	75	0.0019
	100	0.001

Table 9. Relative errors of the pseudo second-order model

T (°C)	time (s)	$(n_r/n_\infty)_{exp}$	$(n_r/n_\infty)_{th}$	$\| [(n_r/n_\infty)_{exp} - (n_r/n_\infty)]/(n_r/n_\infty)_{exp} \|$
30	10	0,00385	0.02029	0.6
	1063	0.69509	0.58632	0,3
	1809	0,95237	0.78988	0,2
50	10	0,00435	0.01459	0.7
	1005	0,66514	0.650338	0.24
	2000	0,73007	0.65366	0.2
75	10	0,00385	0.00454	0.7
	1001	0,35434	0.31456	0.6
	2000	0,38696	0.47722	0.5
100	10	0,00436	0.00791	0.5
	500	0,33112	0.28731	0,5
	1000	0,38996	0.44391	0,3

4.3.5. Fick's Law Model Adjustment Curves

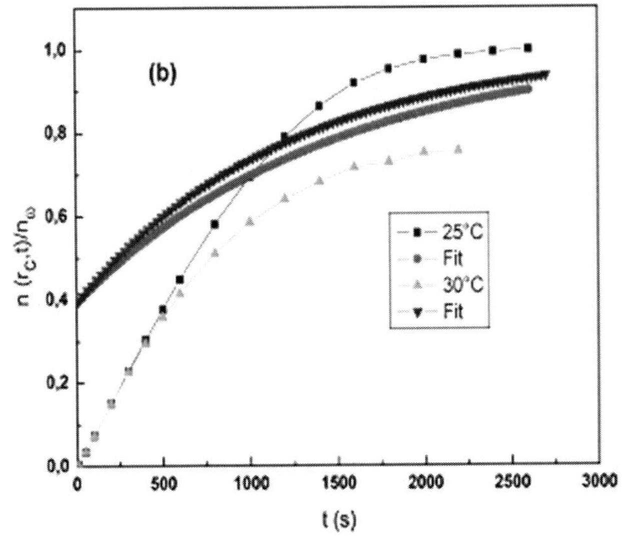

Figure 9. Fick's Law Model Adjustment Curves Acetone.

In this section, we represent the kinetic isotherms relative to Fick's law and the diffusion phenomenon only. It is clear from the figures that the difference between theory and practice is very large for low concentrations and large for concentrations near saturation. As examples, the adjustment curves are shown only for two temperatures.

The following tables list the adjustment results for each pair and the values of the relative errors.

Table 10. Parameter adjustment of Fick's law model

Pair	Temperature (°C)	$D(m^2\ s^{-1})$
Activated carbon-acetone	30	$1.719\ 10^{-11}$
	50	$1.722\ 10^{-11}$
	75	$1.812\ 10^{-11}$
	100	$1.882\ 10^{-11}$

Table 11. Relative errors of the model of Fick's law

T (°C)	time (s)	$(n_r/n_\infty)_{exp}$	$(n_r / n_\infty)_{th}$	$\mid [(n_r/n_\infty)_{exp} - (n_r/n_\infty)] / (n_r/ n_\infty)_{exp} \mid$
30	10	0,00385	0,3956	2.1
	1063	0.69509	0,70873	0,9
	1809	0,95237	0,82841	0,5
50	10	0,00435	0,3956	2.02
	1005	0,66514	0,73859	0.9
	2000	0,73007	0.88418	0.6

4.3.6. Model Fit Curves Based on Improvement of Fick's Law

Below, the experimental and theoretical isotherms for each temperature and each pair are shown. The fit shows a very good match between the experiment and the theory proposed by the model based on the improvement of Fick's law.

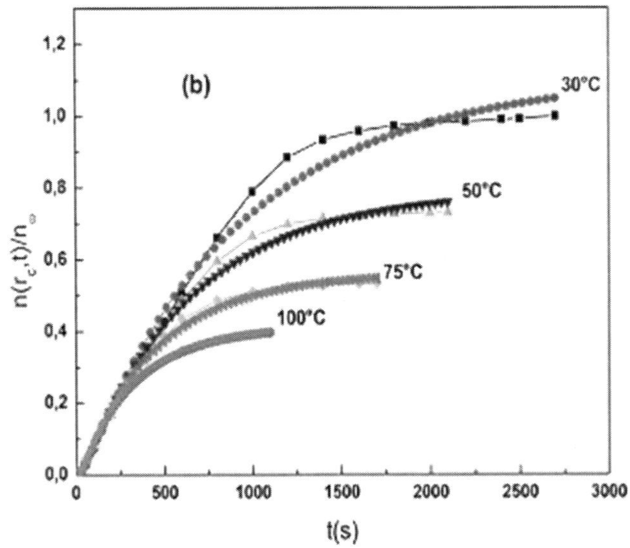

Figure 10. Model fit curves based on the modification of Fick's Law: Acetone.

In this section, the numerical values, adjustment results and relative errors of the model for the different couples studied are tabulated.

Table 12. Acetone Adjustment Result on Activated Carbon

T (°C)	ω^2 **(s^{-1})**	α **(m^{-1})**	τ **(s)**	**δ (without unity)**
30	0.0011	0.6262	145.37	0.3948
50	0.0015	0.838	80.92	0.4228
75	0.0021	1.034	68.241	0.5395
100	0.0029	1.302	53.514	0.8324

The values of the adjustment coefficients given in Table 12 show that the model based on the improvement of Fick's law (Model 4) has values of R2 very close to 1 which explain that this model is in good correlation with the experimental data. Thus, it gives us parameters that are physically acceptable and the relative errors of this model are very low. It does not exceed 10% except for times inferior to 10 s. This is logical, since the initial velocity is very large; the measurements made in the interval [0-10s] are unreliable. But according to the fit curves of the other models, it has been shown that the chemical kinetics is not able to describe the phenomenon since the theoretical equations are in bad correlation with the experimental one and also we notice that the relative errors are very big. Also, these models are only applicable in a certain range of concentrations. This confirms the validity of the model based on physical kinetics.

We found that the adjustment of the experimental adsorption isotherms with the established kinetic models has shown the validity of the kinetic method thanks to the coefficients of determination or adjustment R^2 which are relatively high.

Table 13. Relative errors of the model based on the improvement of Fick's law

T (°C)	time (s)	$(n_r/n_\infty)_{exp}$	$(n_r / n_\infty)_{th}$	$\mid[(n_r/n_\infty)_{exp} - (n_r/n_\infty)] / (n_r/n_\infty)_{exp}\mid$
30	10	0,00385	0,00258	0,3
	1063	0.69509	0,68824	0,009
	1809	0,95237	0,92439	0,02
50	10	0,00435	0,00339	0,2
	1005	0,66514	0,65311	0,01
	2000	0,73007	0,73492	0,006
75	10	0,00385	0,00273	0,2
	1001	0,35434	0,35341	0,002
	2000	0,38696	0,38426	0,006
100	10	0,00436	0,00293	0,3
	500	0,33112	0,32717	0,01
	1000	0,38996	0,38778	0,005

So, we can use the fourth model as an appropriate model to interpret our results.

The fit of the curves by the analytical expression of the best model for all temperatures is shown in Figure (10).

4.4. Interpretation

In the first part of this chapter, we have adjusted the experimental data with the theoretical models to determine the appropriate model. The latter gives us information on the relaxation time relative to the quantity of adsorbed material, minimum interaction distance (adsorbate/adsorbent), relaxation time relative to the fluctuation-dissipation, critical temperature and diffusion constant. The model based on the modification of Fick's law allowed us to better understand the adsorption mechanism by highlighting:

- Relaxation time relative to the quantity of adsorbed material;
- existence of a minimum distance of interaction;
- relaxation time relative to the fluctuation-dissipation of energy and the highlighting of a critical temperature relative to each couple.

4.4.1. The Constant ω^2

$1/\omega^2$ represents the relaxation time relative to the quantity of adsorbed material. This parameter varies according to the couple, which highlights the relative nature of the pairs, given that the transfer of matter into a solid accounts for the properties of the microporous solid that is penetrated and traversed by gaseous molecules from the surrounding medium. The transfer mechanism in microporous solid materials follows an adsorption-diffusion mechanism that consists of a three-step process. The physical adsorption of the gas is generally due to very fast phenomena, essentially transport. On the other hand, diffusion of gas through the solid is very slow. Often, it is this step which imposes the kinetics of the process (limiting step). So, by using this parameter, we can know the dominant phenomenon. For example, when the relaxation time is strong, we talk about diffusion since this phenomenon is a slow one, while the relaxation time is low for adsorption because this phenomenon is fast [39, 40].

The relaxation time relative to adsorption varies as a function of the rate of evolution of heat dissipated during the adsorption process (exothermic phenomenon). When the heat transfer rate is low, the resistance to heat transfer is great. This resistance depends essentially on the thermal conductivity of the torque [38]. In case of the adsorption of gas by a solid, the decrease of conductivity is mainly due to the adsorbate since the lowest thermal conductivity is that of the gas. This conductivity has a very important effect on heat dissipation outside. This conductivity, which blocks heat dissipation to the outside environment and is responsible for the high value of thermal resistance, and consequently of ω^2: Therefore, ω^2 and thermal conductivity vary in the same way. According to the basic kinetic theory of gases, it is shown that the relative variation of conductivity is proportional to the relative variation of temperature. So, we will use this same law to determine the expression of ω^2.

The differential equation satisfying the variation of ω^2 as a function of temperature (equation 22) [38] and integration gives (equation 23).

The variation of ω^2 as a function of temperature (Figure 11) for the pair (acetone-activated carbon) has two asymptotes. These two asymptotes meet at a particular temperature, it is considered a critical temperature T_c, which is an internal parameter of the couple. It is noted that the four isotherms of the studied couples show the existence of the critical temperature. The value of critical temperature is given in Table 15 for the acetone-activated carbon Thus, the origin of this temperature is due to the variation of the physicochemical properties of the adsorbent. If the temperature T_c increases, the loss in mass increases, and the specific surface decreases [41].

Table 15. Estimation of the results of T_c, ω^2_c and f

Pair	ω_c^2 (s^{-1})	T_c (°C)	f (without unity)
Activated carbon-acetone	2.1 10^{-3}	75	0.092

Since the pair is characterized by an internal greatness which is its resistance to mass transfer; this resistance changes of value around T_c. In other words, the resistance to the transfer of the material can take two values: one value for $T < T_c$ and the other for $T > T_c$. That is, there are two regimes, one for $T < T_c$ and one for $T > T_c$.

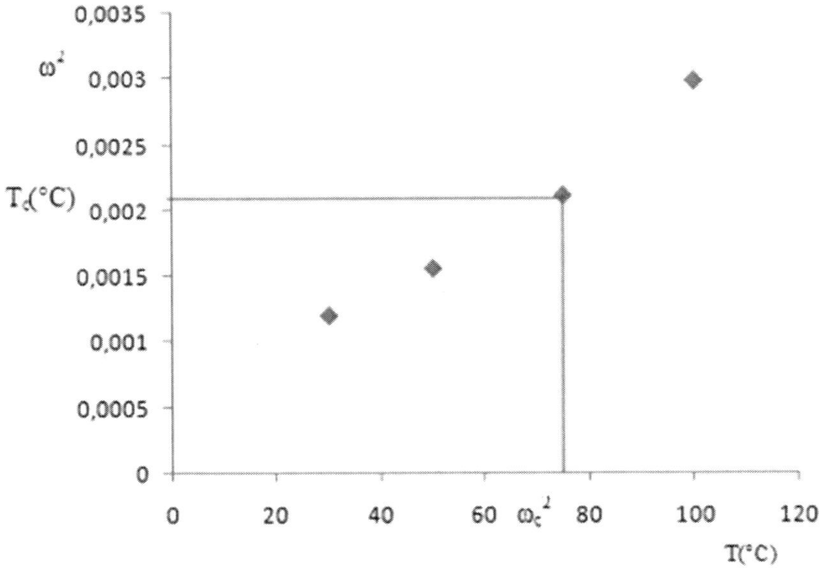

Figure 11. Temperature-dependent evolution: Acetone - activated carbon.

4.4.2. Variation of α^2 as a Function of ω^2

The resolution of the equation (6) made it possible to establish the relation between α and ω^2 in the following form [13]:

$$\alpha = \left(\frac{b + \omega^2}{D} \right)^{1/2} \tag{29}$$

Theoretically, the variation of α^2 as a function of ω^2 is a line of slope 1/D and abscissa at the origin b/D (Figure 12).

Table 16. Estimation of results

Pair	$b(s^{-1})$	$D\ (m^2s^{-1})$
Activated carbon-acetone	$-7.9\ 10^{-4}$	$3.51\ 10^{-3}$

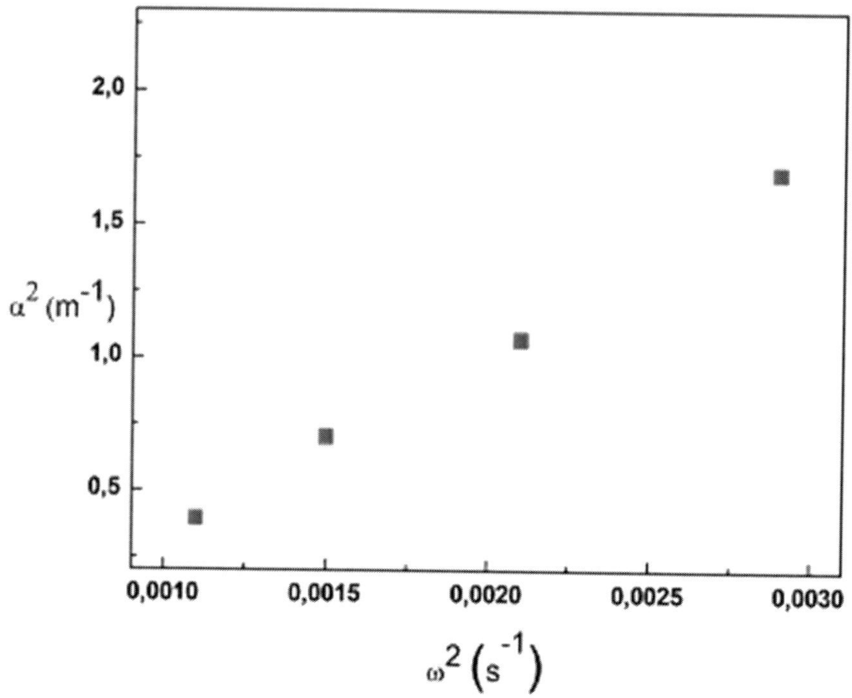

Figure 12. Variation of α^2 as a function of ω^2: Acetone - activated carbon.

Note that the greater the minimum adsorbate-adsorbent interaction distance is, the more important the relaxation time.

4.5. Constant α

$1/\alpha$ represents the minimum interaction distance of the adsorbate-adsorbent pair expressed in meters. Note that α varies as a function of the torque and the temperature, which justifies the relative nature of this magnitude to the couple. As we notice that $1/\alpha$ decreases as temperature increases (Figure 13).

This decrease is explained by the fact that the rise in temperature prevents the adsorbate molecules from easily attaching to the adsorbent surface because of the thermal agitation, that is, when temperature increases. The interactions between the adsorbent surface and adsorbate become increasingly weak under the effect of thermal agitation. Increased collisions prevent weak Van Der Waals or hydrogen bonding from occurring at the receptor site.

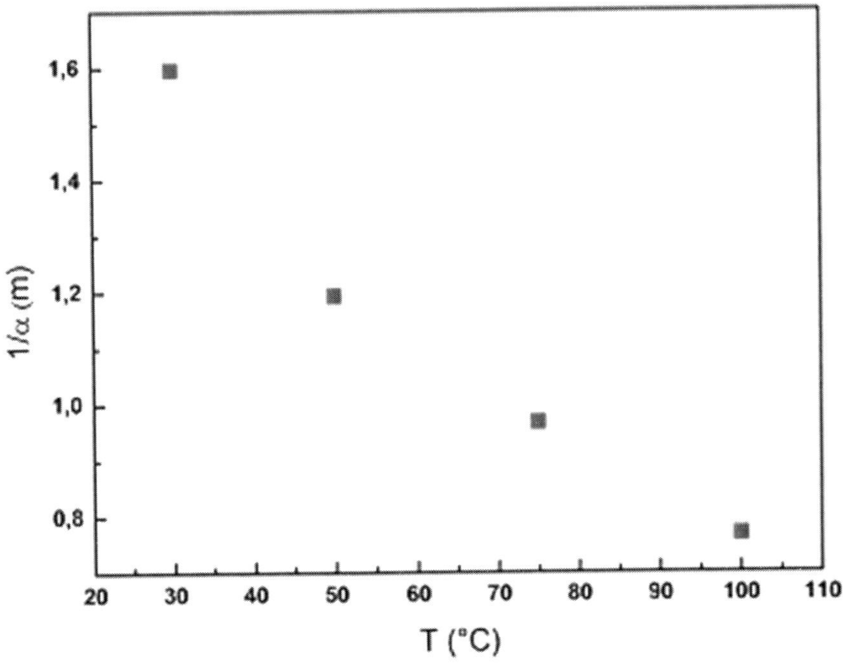

Figure 13. Evolution of 1/α as a function of temperature: Acetone - activated carbon.

4.6. Variation of τ as a Function of Temperature

τ represents the relaxation time relative to fluctuation - dissipation. The variation of τ as a function of temperature for the couple (acetone activated carbon) (Figure 14) shows that the latter decreases when temperature increases.

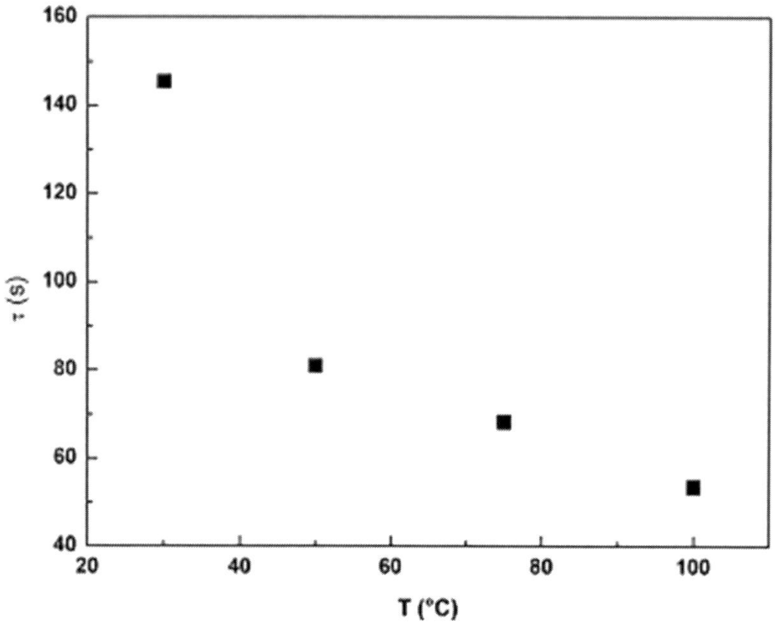

Figure 14. Variation of τ as a function of the torque temperature: Acetone - activated carbon.

At the beginning of adsorption (weak t), the number of empty sites is large, the molecules move easily, which makes the fluctuation of the material important and the relaxation time relative to the fluctuation high. But at the end of adsorption (t large), the adsorbent solid is almost saturated and all the sites are occupied that's why the free movement of the molecules is very low, which means that the amount of fluctuating material is small. So, τ is almost constant.

Conclusion

The benefit of the proposed theory is that it takes into account the non-homogeneity of the system and the fluctuations observed experimentally. The proposed theory takes into consideration: the fluctuation of the adsorbed matter quantity and the energy fluctuation; the fluctuation of the adsorbed matter quantity and the dissipation of the thermal energy.

This theory takes into account all the internal system values because of the use of the statistical theory and the random function while solving the established differential equation.

From the previous studies and the present one, we can affirm that the adsorption of a gas in a microporous solid is the result of three phenomena: a transportation phenomenon followed by a diffusion phenomenon and both phenomena are accompanied by a fluctuation.

In spite of the assumption that the adsorption is isobar, the concordance between the theory and the experiment is very satisfactory. This is attributed to the fact that the studied adsorption is free and not forced, and the pressure variations are very small and can be neglected. The relative error between practice and theory is less than 10% at any instant over 10s. This result is expected as the initial speed is very high and the measurements during the time interval [0, 10s] are less assured.

REFERENCES

[1] Crank, J. *Mathématics of diffusion*; Oxfor Uni. Press. London G.B. (1956). Chapter VI.

[2] Vavlitis, A.P., Ruthven, D.M and Loughlin, K.F. 1981. Sci.84: 526-531. Sorption of N-Pentane, N-Octane and N-Decane in 5A zeolite crystals. *Colloid.Interface. J.*

[3] Kondis, E.F., and Dranoff, J.S. 1971. 10: 108-114. Kinetics of isothermal sorption of Ethane on 4A molecular sieve pellets. *Ind. Eng. Chem.* Process Design Develop.

[4] Barrer, R.M., and Gibbons, M., 1965. 61: 948-961.Zeolitie carbon dioxide: Energetics and equilibria in relation to exchangeable cations in Faujasit. *Trans. Far. Soc.*

[5] Barrer, R.M. 1984. *Zéolithe Sciences and Technology*. Nato ASI SER. 227-249].

[6] Fowkes, F.M. 1967. *Surfaces and interfaces I.* Syracuse Univ. Press, 197-223.

[7] Habgood, H.W. (1967). 63, 45-51. "Adsorption onto heterogeneous porous materials: Equilibria and Kinetics" - *Chem. Eng. Symp. Ser. Prog.*

[8] Prigogine, I., 1955. *Introduction to thermodynamics of Irreversible Processes*. Charles C. Tomas, Springfield, IL. US.

[9] Nicolis, G., and Prigogine, I. 1977. *Self-Organisation in Nonequilibruim Dystems. From Dissipative Structures to Order through Fluctuation*, John Wiley & Sons. New York/London/ Sydney/Toronto.

[10] Onsager, L. 1931. 37: 405-426. Reciprocal relation in irreversible processes. *Phys. Rev.*

[11] Onsager, L. 1931. 38 : 2265-2279. Reciprocal relation in irreversible processes. *Phys. Rev.*

[12] Mhiri, F., Jemni, A., El golli, S. 1996. 35 : 39-47. Etude de l'énergie de désorption des substances polaires. *Rev. Gén. Therm.*

[13] Kazymyr P., Lyubov P., Oleg T. 2000. 18-1:15-26 Oscillatory adsorption as the determination of the fluctuating behaviour of different heterogeneous systems. *Adsorption Science & Technology.*

[14] Czarniecki, J., and Jaroniec, M. 1983. 3-6: 301-353. Studies of adsorption kinetics by means of the stochastic numerical simulation. *Surface Science Reports.*

[15] Lozano-Castelló, D., Maciá-Agulló, J.A., Cazorla-Amorós, D., Linares-Solano, A., Müller, M., Burghammer, M., Riekel, C. 2006. 44-7:1121-1129. *Isotropic and anisotropic microporosity development upon chemical activation of carbon fibers, revealed by microbeam small-angle X-ray scattering*. Carbon.

[16] Perre-Henri C. 2006. 36-1: 81-123. Hamiltonian and Brownian systems with long-range interactions: II. Kinetic equation and stability analysis, *Physica A.*

[17] McEwen J. S., and al., 2003.545: 47-69. Adsorption and desorption of CO on Pt(111):a comprehensive analysis, *Surface Science.*

[18] Fan X., and al., 2003. 37: 4929-4937. Adsorption kinetic of fluoride on low cost materials. *Water Research.*

[19] Javier Rivas F. and al., 2003. 37: 4823-4834. Stabilized leachates: ozone-activated carbon treatment and kinetics. *Water Research.*

[20] Jingtian Hu and al. 2003. 96: 390-398. Estimation the kinetics parameters for non-specific adsorption of fibrinogen on quartz surface from the response of an electrode-separated piezoelectric sensor. *Sensors and actuators B.*

[21] Mikko S. and al. 2003. 259: 73-81. Liquid-phase hydrogenation kinetics of isoctenes on Co/Si_2. *Applied Catalysis A.*

[22] Nitin K. and al. 2003. 267: 272-285. Measurement of the kinetic rate constants for the adsorption of superspreading trisiloxanes to an air/aqueous interface and the relevance of these measurements to the mechanisim of superspreading. *Journal of Colloid and Interface Science.*

[23] Budinski- Petkovie Ljand Tosie T. 2003. 329: 350-356. Adsorption, desorption and diffusion of extended objects on a square lattice. *Physica A.*

[24] Xiaojun C. and al. 2004. 83:293-303. Selective transport of CO_2, CH_4, and N_2 in coals: insights from modelling of experimental gas adsorption data. *Fuel.*

[25] Joaquin C. and Eliana V. 2003. 267: 391-396. Structural effects in the kinetic behaviour of the monomer-dimer surface reaction over nondeterministic fractal surfaces. *Journal of Colloid and Interface Science.*

[26] Marcelo A. and Edward J. F. 1977. General Physics, Tomme II. *Fields and Waves*; Inter Editions, Paris, chapter 24.

[27] Nikiforov A. and Ouvarov V. 1983. *Special Functions of Mathematical Physics*, Edition MIR. Mouscou, chapter 1, French translation 1983.

[28] Blanc-Lapierre A. and Picinbono B. 1981. *Random Functions*, Masson Edition (1981). Chp. 4.

[29] Aïvazian S. 1978. *Statistical study of dependencies*, MIR Editions. Moscow (Second Edition), Chapters II, IV, V, VI.

[30] Mhiri F., Jemni A., Ben Nasrallah S. 2009. 12: 563-571. Kinetic adsorption of water and carbon Dioxide in zeolites. *Journal of Porous Media.*

[31] Mhiri F., Jemni A. 2010. 13: 295-305. Kinetic study for the adsorption of acetone and ethanol on activated carbon. *Journal of Porous Media.*

[32] Mhiri F. and Jemni A. 2011. 30-3: 294-302. Correlation between fluctuation and dissipation in the case of the adsorption of acetone and ethanol by activated carbon. *Environmental Progress & Sustainable Energy.*

[33] Arthur W. A. (1990). Chp. XVII. *Physical chemistry of surfaces* (fifth edition). John Wiley & Sons, Inc. (1990). Chapter XVII.

[34] Nahid, G., Sara, M., Maryam, G., Somaye, M., Mohammad, H. T. 2012. 3. Adsorption isotherms and kinetics Studies for the removal of Pb(II) from Aqueous Solutions using Low-cost adsorbent. *International Conference on Environment Science and Engineering*, Singapore.

[35] Ibrahim I. El-S., Kutub, U., Takahiko, M., Bidyut, B. S., Shigeru K., Jin M., Seong-Ho, Y. 2014. 73: 445-455. Adsorption of ethanol onto parent and surface treated activated carbon Powders. *International Journal of Heat and Mass Transfer.*

[36] Diu, B., Guthmann, C., Lederer, D., B. Roulet, B. 1989. *Statistical physics.* Hermann. Paris 1ere edition.

[37] Terzyk, A.P., and Rychlicki, G. 1999. 17: 87-100. Calorimetric Investigation of Molecular Interaction in the Adsorbate/ Microporous Activated Carbon System. Towards the Mechanism of Adsorption in Micropores. *Adsorption science and Technology.*

[38] Gregg. S.J. and Sing, K.S.W. 1982. 19-107. *Adsorption, Surface Area and Porosity.* Academic Press. London. Hagiwara, S., Tsutsumi, K. and Takahashi, H. Carbon.

[39] Atkinson, D., McLeod, A.I. and Sing, K.S.W.1984. 81-791. *Journal of Chemical.Physics.*

In: Acetone
Editors: Irenka Kozłowska et al.
ISBN: 978-1-53613-298-4

Chapter 2

LYSOZYME IN WATER-ACETONE MIXTURES: RESIDUAL ENZYME ACTIVITY AND PREFERENTIAL SOLVATION

Vladimir A. Sirotkin ***and Alexandra A. Kuchierskaya***
A. M. Butlerov Institute of Chemistry, Kazan Federal University, Kazan, Russia

ABSTRACT

Numerous benefits of employing enzymes in non-aqueous organic liquids exist, including catalysis of industrially-important synthetic reactions, suppression of undesirable side reactions caused by water, and high solubility of hydrophobic reagents. Residual enzyme activity was investigated to monitor stabilization/destabilization of hen egg-white lysozyme at low, intermediate, and high water content in acetone at 25°C. The advantages of our methods are as follows:

1. Residual enzyme activity curves can be determined in the entire range of water content in organic liquids.
2. Enzyme activity values can be measured at fixed reaction conditions.

* Corresponding Author Email: vsir@mail.ru.

3. Information about changes in the state of the catalytically-active sites can be obtained separately from other contributions (e.g., changes in the solvation of reagents and products in water – organic mixtures).

Our results clearly demonstrate that the stabilization/destabilization of lysozyme depends significantly on water content in acetone.

1. At high water content, residual activity values are close to 100%.
2. A minimum on the residual activity curve was observed at intermediate water content. Acetone augments the irreversible inactivation of lysozyme in this region.
3. Lysozyme demonstrates significant residual activity in water-poor acetone.

Keywords: biocatalysis in water-organic mixtures, protein stability, preferential solvation, enzyme activity, lysozyme; acetone

1. INTRODUCTION

Enzyme-water interactions are well-known to play a critical role in determining the function, structure, and stability of enzyme macromolecules [1-8]. Elucidation of the processes occurring upon enzyme hydration in the presence of organic solvents is essential in a wide range of biomedical and biotechnological applications. In particular, organic solvents are widely utilized as model substances to induce amyloid fibril and molten globule formation [9, 10]. Numerous debilitating diseases are associated with these protein states, including type II diabetes, Parkinson's disease, and Alzheimer's disease.

Enzyme activity constitutes an intricate function of water content in organic solvents. Typical dependencies of enzymatic activity on water content in organic solvents can be delineated into three parts [11-15]:

1. The first range of concentration concerns water-rich mixtures. In this range, one can observe hydrolytic enzyme activity. However, numerous reactions that are vital to industry, such as esterification

and peptide synthesis, are suppressed in aqueous solutions as a consequence of the unfavorable shift of reaction equilibria.

2. A sharp decline in enzyme activity was found after a certain concentration threshold of the organic solvent had been reached. The physicochemical properties of the solvent determine the position of this minimum [13]. The enzyme structure may be disturbed by the organic solvents through weakening of the hydrophobic interactions, through altering the electrostatic interactions of the polar protein groups, and/or by direct interaction with the biocatalysts.
3. Mixtures with low water content constitute the third concentration range. The dehydrated enzymes are in a kinetically "frozen" state [1, 6, 7, 14-23]. Due to the reduced conformational flexibility in the water-poor organic solvents, the enzymes remain in the active conformation. There are some benefits of employing enzymes in low-water organic liquids (e.g., high solubility of hydrophobic substrates, favorable shift of thermodynamic equilibrium in the synthetic direction [peptide synthesis and transesterification] versus hydrolysis, etc.) [11-23].

Preferential solvation is an effective and informative method for elucidating the mechanism of protein stabilization or denaturation [24-39]. When a protein interacts with a binary water-organic solvent mixture, the three components do not mix equally. Water or organic solvent molecules exist preferentially in the protein's solvation shell. This difference between the solvation shell and bulk solvent in the solvent components has been termed *preferential solvation* [24-26, 32, 33, 38]. Preferential solvation is a thermodynamic quantity that describes the protein surface occupancy by the water and organic solvent molecules. This is associated with the actual numbers of water/organic solvent molecules that are in contact with the protein's surface. It was also found that protein destabilization is directly associated with preferential binding of denaturant molecules to specific protein groups [24-27, 29, 32, 33, 36].

Preferential hydration constitutes an excess of water in the protein macromolecule relative to water concentration in the overall solvent. Preferential hydration does not always stabilize native proteins [33]. Preferentially hydrating solvent systems can be divided into two groups:

1. The first always stabilizes the protein structure. The dominant interaction in the first group is organic solvent exclusion, in which the protein remains essentially inert.
2. Preferential interactions in the second group are governed by the protein surface's chemical nature. This achieves a precise balance between the binding and exclusion of organic solvent. Exposure of the additional protein hydrophobic groups can be enhanced by protein unfolding.

Preferential interactions can be measured by a variety of techniques: differential refractometry and densitometry [37, 2, 38], NMR spectroscopy [39], isopiestic measurements of vapor pressure [36], etc. These works were performed at high water content. No attempt has yet been made, however, to investigate preferential solvation of protein macromolecules at low- and high water content in organic solvents simultaneously.

The aim of our study is to monitor preferential hydration and preferential solvation in the entire range of water content in water-organic mixtures at 25°C. Our approach is based on simultaneous measurements of the absolute values of water and organic solvent sorption. Preferential solvation/hydration parameters were calculated using the water and organic solvent sorption values. The advantages of our approach are as follows:

1. Our approach facilitates individual evaluation of the Gibbs energies of protein, water, and organic solvent;
2. Preferential interaction parameters can be determined in the entire range of water content in organic liquids.

Hen egg-white lysozyme was used as a model protein. This protein is one of the most applied and studied in biotechnological and biothermodynamic research [40, 41]. Lysozyme is a small monomeric protein of 129 amino acid residues. The physiological role of lysozyme is to hydrolyze polysaccharide chains [40, 41]. Acetone was chosen for the following reasons:

1. This solvent is widely utilized in enzymology and biotechnology [13, 16, 17, 19];
2. Acetone is an organic solvent that is water-miscible. Consequently, the effect of this organic molecule on the hydration and structure of lysozyme in the entire range of water content can be investigated;
3. Acetone is able to form hydrogen bonds with several hydrogen donors. In contrast to water, however, it does not possess hydrogen bond-donating ability.

2. EXPERIMENTAL

2.1. Materials

Hen egg-white lysozyme (EC 3.2.1.17; crystallized three times, dialyzed, and lyophilized and dried *Micrococcus lysodeikticus* cells were purchased from Sigma Chemical Co. (St. Louis, MO, U.S.A.). The molecular weight of the protein was taken as 14300 Da. Acetone (analytical grade, with a purity of >99%) was purified and dried according to the recommended guidelines [42]. Water used was doubly distilled. All water-organic mixtures were prepared gravimetrically using a Precisa balance (Swiss) with a precision of 0.00001 g.

2.2. Zero Hydration Level

Dried protein preparation was obtained through vacuum drying utilizing a microthermoanalyzer "Setaram" MGDTD-17S (±0.00001 g) at

25°C and 0.1 Pa, until obtaining a constant sample weight. The dried protein's water content was estimated as 0.002 ± 0.001 g water g^{-1} protein using the Karl Fischer titration method, according to the recommended guidelines [43, 44]. This value for lysozyme implies that there exist approximately three water molecules strongly bound to each protein molecule at the zero hydration level.

2.3. Sorption Measurements

The protein samples were prepared as described previously [45-48]. The samples of initially dehydrated protein were presented to water-acetone vapor mixtures. The water-acetonitrile vapor mixture was flowed consecutively through a thermostated saturator filled with the water-organic mixture, and a cell containing the protein sample. Protein samples (8-10 mg) were each flushed by water – organic vapour mixtures until no further mass changes were detected, as described previously [46]. Typically, sorption equilibrium was reached after 6 h at 25°C. Measurements of the protein-bound water (A_1) were performed by Karl Fischer titration using a Metrohm 831 KF coulometer. This constitutes a well-known method to determine the water content of protein powders. Acetone content of lysozyme (A_3) was calculated as the difference between the total sorption uptake (A_1+A_3) and water content (A_1). The total sorption uptake (A_1+A_3) was measured by microthermoanalyzer "Setaram" MGDTD-17S. Water activity (a_1) in the vapor phase was adjusted by altering the aqueous-organic mixture within the saturator.

2.4. Residual Enzyme Activity

Residual enzyme activity was determined by measuring the enzyme activity after storage in water-organic mixtures, as described previously

[46]. Lysozyme activity was determined as follows. Lysozyme was immersed in an aqueous-organic mixture of required composition, and incubated at 25°C for 3 h. This time period exceeded the time corresponding to the completion of the calorimetric heat effect accompanying the interaction of the dehydrated proteins with pure organic solvents and water-organic mixtures [49].

The concentration of lysozyme in the water-organic mixtures was 1 mg/ml. Adding 100-μl aliquots of the lysozyme solution in water-acetone mixtures to the aqueous solution of the substrate (*Micrococcus lysodeikticus* cells [2.9 ml, 0.3 mg/ml] in 0.1 M potassium phosphate pH 7.0), we initiated the enzymatic reaction. Change in the absorbance at 450 nm was recorded using a Perkin-Elmer Lambda 35 double-beam scanning spectrophotometer. The reaction was followed for 300 - 1500 s. Each kinetic curve was reproduced at least three times.

3. Results and Discussion

3.1. Sorption Isotherms and Excess Sorption Functions

Figure 1 shows the water (A_1) and acetone (A_3) vapor sorption isotherms for lysozyme at 25°C. The protein solvation shell comprises two parts: (1) non-ideal (due to preferential solvation and preferential hydration); and (2) ideal.

The non-ideal effect of the solvation shell on enzyme properties (residual enzyme activity, acetone/water sorption) can be articulated in terms of excess functions, F^E [50-52], which refers to the difference between the observed function of mixing, F^M, and the function for an ideal binary mixture, F^M_{id}.

The deviations of the excess functions from zero describe the extent to which the protein solvation shell differs from the pure binary water-organic system as a consequence of preferential interactions occurring between water (component 1), protein (component 2), and acetone (component 3).

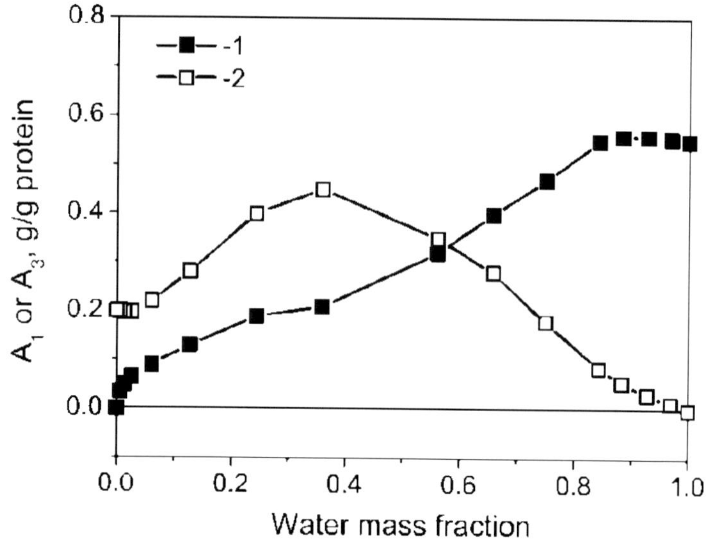

Figure 1. Water (A_1) sorption isotherm for lysozyme at 25°C. 2 - Acetone (A_3) sorption isotherm for lysozyme at 25°C. The standard errors of estimation of the water/acetone sorption were 0.001 - 0.002 g/g. Each experiment was performed 3 - 4 times.

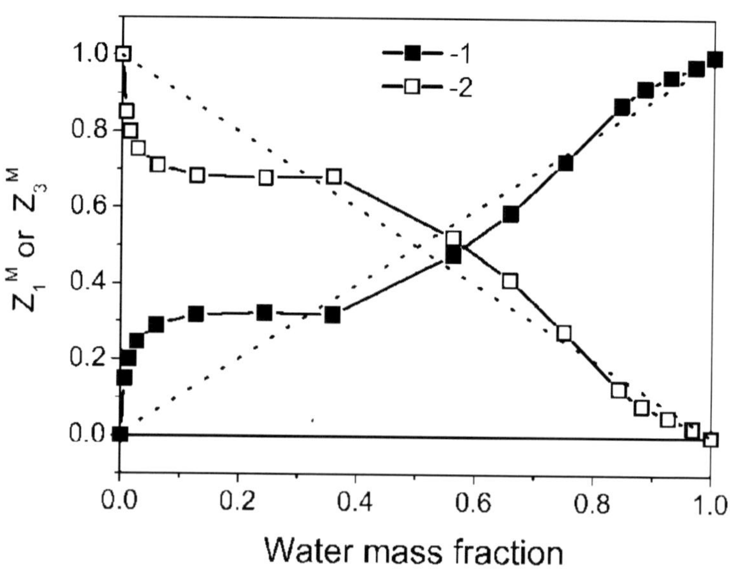

Figure 2. Water mass fraction in the solvation layer of lysozyme (Z_1^M). 2 - Acetone mass fraction in the solvation layer of lysozyme (Z_3^M).

Eq. 1 is used to calculate the F^E values:

$$F^E = F^M - F_{id}^M \quad (1)$$

The $F_{id,i}^M$ values can be calculated using Eq. 2:

$$F_{id,i}^M = F_i^M(w_i = 0) + w_i[F_i^M(w_i = 1.0) - F_i^M(w_i = 0)] \quad (2)$$

where $F_i^M(w_i = 1.0)$ is the observed mixing function of lysozyme at $w_i = 1.0$; $F_i^M(w_i = 0)$ is the observed mixing function of lysozyme at $w_i = 0$; w_1 is the water mass fraction in the binary water-organic mixtures; and w_3 is the organic solvent mass fraction in the binary water-organic mixtures ($w_1 + w_3 = 1.0$).

The situation in which no preferential interactions exist between water, protein, and acetone is described by the $F_{id,i}^M$ values. In this circumstance, the water mass fraction in the ideal part of the solvation shell is the same as in the pure water-organic mixture.

The Z_1^M (water mass fraction in the solvation shell) and Z_3^M (organic solvent mass fraction in the solvation shell) values as a function of water mass fraction in acetone are shown in Figure 2. Eqs. 3 and 4 are used to calculate the Z_1^M and Z_3^M values, respectively:

$$Z_1^M = \frac{A_1}{A_1+A_3} \quad (3)$$

$$Z_3^M = \frac{A_3}{A_1+A_3} \quad (4)$$

The simultaneous action of acetone and water was characterized by the Z_1^E and Z_3^E values (Figure 3). These excess sorption functions were calculated using Eqs. 5 and 6:

$$Z_1^E = Z_1^M - Z_{id,1}^M \quad (5)$$

$$Z_3^E = Z_3^M - Z_{id,3}^M \tag{6}$$

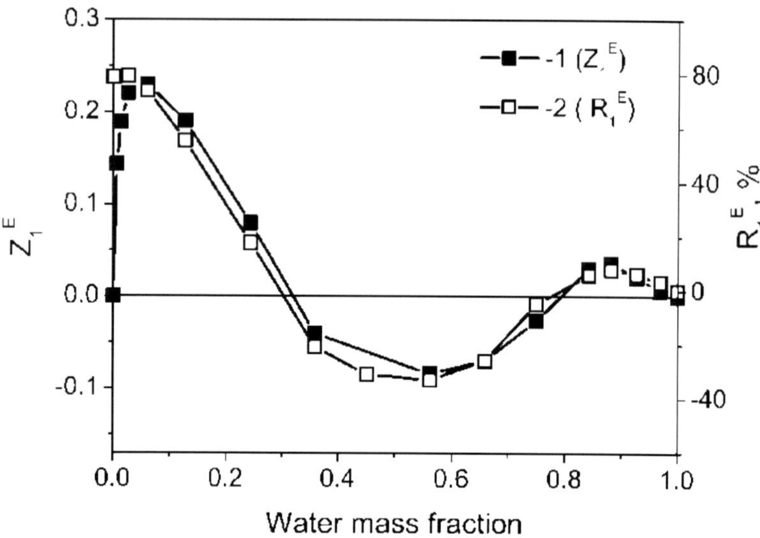

Figure 3. Excess water mass fraction in the solvation layer of lysozyme (Z_1^E).
2 - Excess residual activity of lysozyme in water-acetone mixtures (R_1^E).

where Z_1^M is the mass fraction of water in the solvation layer for the real water-organic mixture; and $Z_{id,1}^M$ is the mass fraction of water in the solvation layer for the ideal water-organic mixture. The $Z_{id,1}^M$ values were calculated using Eq. 7:

$$Z_{id,1}^M = Z_1^M(w_1 = 0) + w_1[Z_1^M(w_1 = 1.0) - Z_1^M(w_1 = 0)] \tag{7}$$

where $Z_1^M(w_1 = 1.0)$ is the water mass fraction in the solvation shell of lysozyme at w_1=1.0; $Z_1^M(w_1 = 0)$ is the water mass fraction in the solvation shell at w_1=0; and w_1 is the mass fraction of water in organic solvent.

In addition, Z_3^M is the organic solvent mass fraction in the solvation shell for the real water-organic mixture; and $Z_{id,3}^M$ is the organic solvent mass fraction for the ideal water-organic mixture. Eq. 8 is used to calculate the $Z_{id,3}^{M}$ values:

$$Z_{id,3}^{M} = Z_3^M(w_3 = 0) + w_3[Z_3^M(w_3 = 1.0) - Z_3^M(w_3 = 0)] \qquad (8)$$

where $Z_3^M(w_3 = 0)$ is the organic solvent mass fraction in the solvation shell of lysozyme at $w_3=1.0$; $Z_3^M(w_3 = 0)$ is the organic solvent mass fraction in the solvation shell of lysozyme at $w_3=0$; and w_3 is the organic solvent mass fraction in the binary water-organic mixture.

The Z_1^E values are positive at high ($w_1 = 0.8\text{-}1.0$) and low ($w_1 = 0\text{-}0.2$) water content (Figure 3). A considerable decline in water sorption was found in the intermediate range of water content. The Z_1^E values are negative in this concentration region. The most pronounced suppression was observed in the water mass fraction range from 0.4 to 0.8.

3.2. Residual Enzyme Activity

Figure 4 demonstrates typical kinetic curves for the enzymatic reaction catalyzed by lysozyme preliminarily incubated in water-acetone mixtures. The catalytic activity was characterized by the ratio of the extent of hydrolysis attained within 300 s with lysozyme incubated in a water-organic mixture to the same quantity measured using lysozyme incubated in pure water (Figure 4, curve 1).

The residual activity values are presented in Figure 5. As shown in the figure, acetone affects the residual activity of lysozyme in a complicated manner.

The residual enzyme activity values are close to 100% (Figure 3) at high water content ($w_1 \sim 0.9 - 1.0$). There is a sharp transition from the water-rich range to the intermediate one at $w_1 < 0.9$. The residual enzyme activity of lysozyme changes from 100 to 0% in the transition region ($w_1 \sim 0.6 - 0.8$). A minimum on the residual activity curve was found at $w_1 \sim 0.6$. The residual activity augments at $w_1 < 0.4$. At low water content, the residual enzyme activity is ~ 80%, compared with that observed after incubation in pure water.

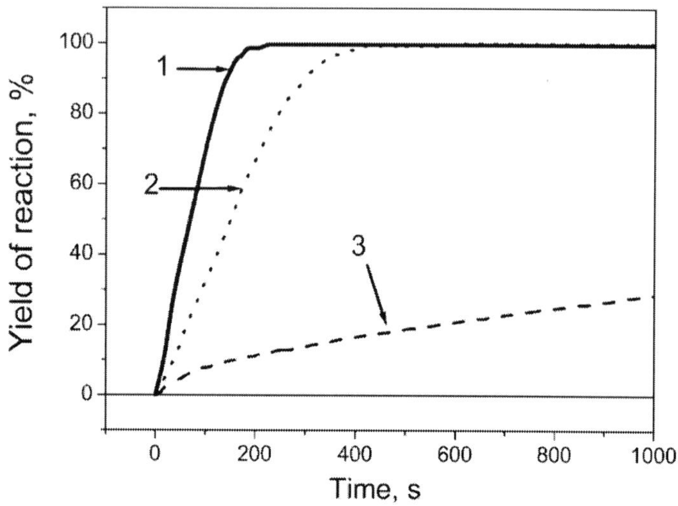

Figure 4. Typical kinetic curves for the enzymatic reaction catalyzed by lysozyme previously incubated in water-acetone mixtures. Water mass fraction in ethanol: (1) 1.0; (2) 0.14; (3) 0.56.

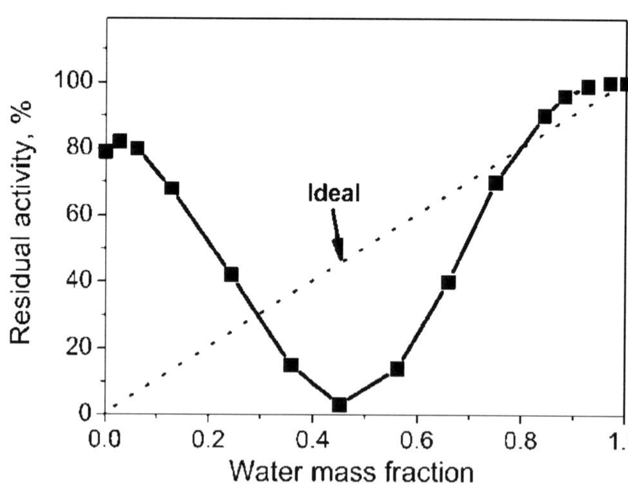

Figure 5. Residual activity of lysozyme in water-acetone mixtures. All values are the averages of three measurements. Experimental errors were 1 - 1.5%.

3.3. Excess Residual Enzyme Activity

The effect of the excess hydration ($\boldsymbol{Z_1^E}$) of the residual enzyme activity was described by the $\boldsymbol{R_1^E}$ values (excess residual enzyme activity) (Figure 3). The $\boldsymbol{R_1^E}$ values were estimated using Eq. 9:

$$R_1^E = R^M - R_{id,1}^M \tag{9}$$

where R^M is the observed residual enzyme activity; and $R_{id,1}^M$ is the function for an ideal binary mixture.

The R_{id}^M values can be calculated using Eq. 10:

$$R_{id,1}^M = R^M(w_1 = 0) + w_1[R^M(w_1 = 1.0) - R^M(w_1 = 0)] \tag{10}$$

where $R^M(w_1 = 1.0)$ is the residual activity of lysozyme at $w_1 = 1.0$; $R^M(w_1 = 0)$ is the residual activity of lysozyme at $w_1 = 0$; w_1 is the water mass fraction in the binary water-alcohol mixtures; and w_3 is the alcohol mass fraction in the binary water-organic mixtures ($w_1 + w_3 = 1.0$).

Figure 3 demonstrates the dependence of the R_1^E values on the water content in acetone. As shown in Figure 3, the R_1^E values differ significantly from zero. This means that the effect of the water-organic solvation shell on the residual enzyme activity is non-ideal in the entire range of water content. It is worth noting that the R_1^E values are consistent with the Z_1^E values (Figure 3).

Three concentration regions exist (Figure 3):

1. At high water content, the R_1^E and Z_1^E values are positive;
2. At intermediate water content, the R_1^E and Z_1^E values are negative. Acetone augments the degree of irreversible inactivation of lysozyme in this region;
3. At low water content, the R_1^E and Z_1^E values are positive.

3.4. Preferential Interaction Parameters

The degree to which the solvation layer differs from the pure binary water-acetone mixture due to preferential interactions between water (component 1), protein (component 2), and acetone (component 3) can be characterized by the preferential interaction parameters (Eqs. 11 and 12). Eq. 11 [38] was utilized to determine the preferential solvation of lysozyme in water-acetone mixtures:

$$(\partial g_3/\partial g_2)_{T,\mu 1,\mu 3} = A_3 - \frac{w_3}{w_1}A_1 \qquad (11)$$

where A_1 is the protein hydration, expressed as gram water per gram protein; A_3 is the binding of acetone, expressed as gram organic solvent per gram protein; w_1 is the water mass fraction in water-acetone mixtures; and w_3 is the mass fraction of organic solvent in water-acetone mixtures ($w_1 + w_3 = 1.0$).

Eq. 12 was used to estimate preferential hydration:

$$(\partial g_1/\partial g_2)_{T,\mu 1,\mu 3} = -(\frac{w_1}{w_3})(\partial g_3/\partial g_2)_{T,\mu 1,\mu 3} \qquad (12)$$

Figures 6 and 8 show the preferential interaction parameters calculated using Eqs. 11 and 12, respectively.

Eqs. 13 and 14 were utilized to determine the Gibbs energies of the transfer of water ($\Delta \boldsymbol{G_1^{pref}}$) and acetone ($\Delta \boldsymbol{G_3^{pref}}$) from water-organic mixtures to the solvation layer of lysozyme:

$$\Delta G_1^{pref} = \mu_1^E(solvation\ layer) - \mu_1^E(binary\ mixture) \qquad (13)$$

$$\Delta G_3^{pref} = \mu_3^E(solvation\ layer) - \mu_3^E(binary\ mixture) \qquad (14)$$

The $\mu_1^E(binary\ mixture)$ and $\mu_3^E(binary\ mixture)$ values can be calculated using Eqs. 15 and 16:

$$\mu_1^E(binary\ mixture) = RTln\gamma_1(binary\ mixture) \quad (15)$$

$$\mu_3^E(binary\ mixture) = RTln\gamma_3(binary\ mixture) \quad (16)$$

Water activity coefficients ($\gamma_1[binary\ mixture)$], the mass fraction scale; the reference state is pure water) in aqueous-organic mixtures were estimated using Eq. 17:

$$\gamma_1(binary\ mixture) = \frac{a_1}{w_1} \quad (17)$$

Acetone activity coefficients ($\gamma_3[binary\ mixture)$], the mass fraction scale; the reference state is pure acetone) in aqueous-organic mixtures were calculated using Eq. 18:

$$\gamma_3(binary\ mixture) = \frac{a_3}{w_3} \quad (18)$$

Water activity (a_1) and acetone activity (a_3) were taken from published data [53,54] based on the vapor-liquid equilibrium.

The $\mu_1^E(solvation\ layer)$ and $\mu_3^E(solvation\ layer)$ values were calculated using Eqs. 19 and 20:

$$\mu_1^E(solvation\ layer) = RTln\gamma_1(solvation\ layer) \quad (19)$$

$$\mu_3^E(solvation\ layer) = RTln\gamma_3(solvation\ layer) \quad (20)$$

Water activity coefficients ($\gamma_1[solvation\ layer)$], the mass fraction scale) in the solvation layer were calculated using Eq. 21:

$$\gamma_1(solvation\ layer) = \frac{a_1}{Z_1^M} \quad (21)$$

where Z_1^M is the mass fraction of water in the solvation layer; $Z_1^M = \frac{A_1}{A_1+A_3}$.

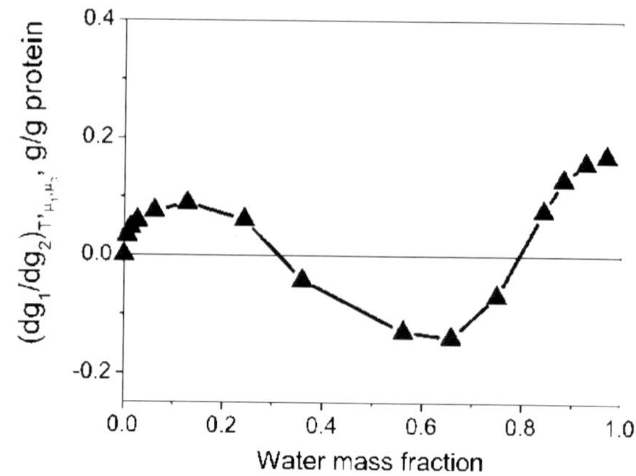

Figure 6. The preferential hydration parameter as a function of water mass fraction in acetone ($[\partial g_1/\partial g_2]_{T,\mu1,\mu3}$).

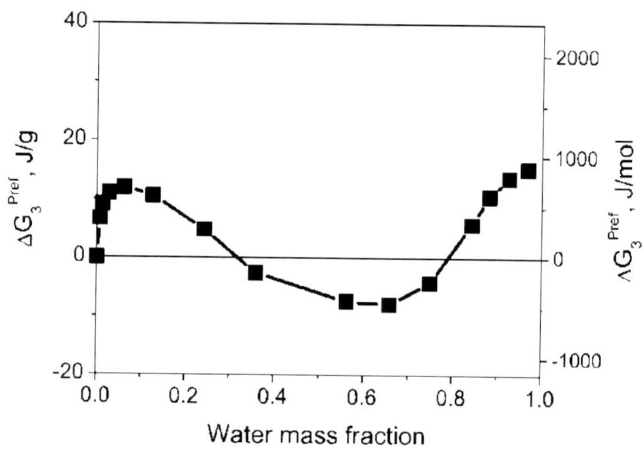

Figure 7. Gibbs energy of the transfer of acetone (ΔG_3^{pref}) from binary water-organic mixtures to the solvation shell of lysozyme.

Acetone activity coefficients ($\gamma_3[solvation\ layer]$, the mass fraction scale) in the solvation layer were estimated using Eq. 22:

$$\gamma_3(solvation\ layer) = \frac{a_3}{Z_3^M} \qquad (22)$$

where Z_3^M is the acetone mass fraction in the solvation layer; $Z_3^M = \frac{A_3}{A_1+A_3}$.

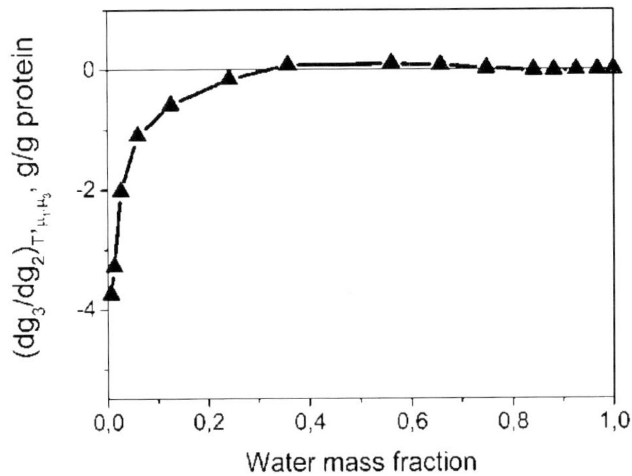

Figure 8. The preferential solvation parameter as a function of water mass fraction in acetone ($[\partial g_3/\partial g_2]_{T,\mu 1,\mu 3}$).

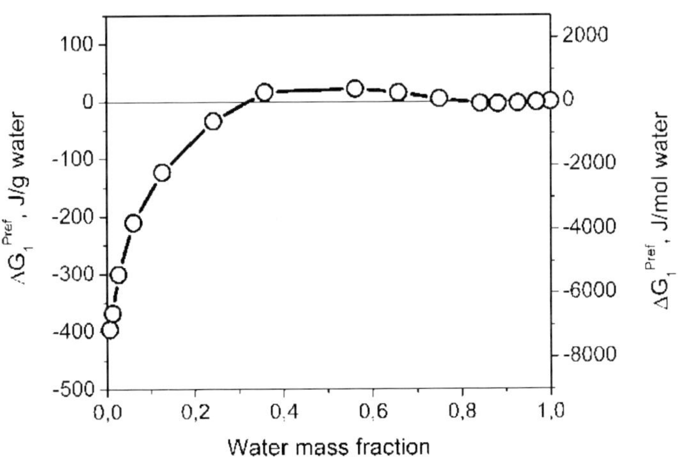

Figure 9. Gibbs energy of the transfer of water (ΔG_1^{pref}) from binary water-organic mixtures to the solvation shell.

Figures 7 and 9 show the ΔG_3^{pref} and ΔG_1^{pref} values, respectively. Figures 6 - 9 demonstrate that the ΔG_3^{pref} and ΔG_1^{pref} values correlate well with the preferential interaction parameters.

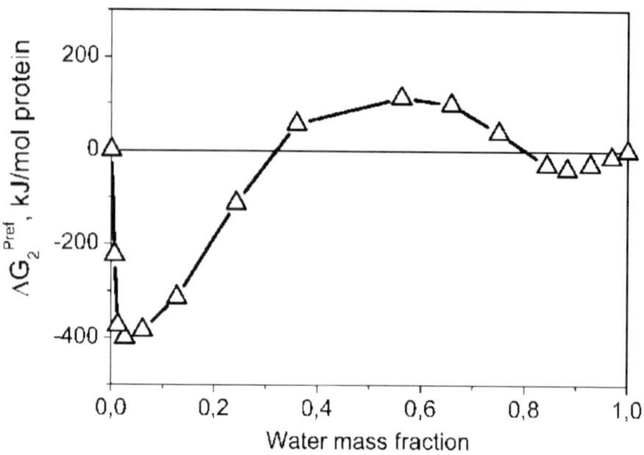

Figure 10. Gibbs energy (ΔG_2^{pref}) of the transfer of lysozyme from pure state to the water-acetone mixtures.

ΔG_2^{pref} values (Figure 10) describe lysozyme stabilization and destabilization due to the preferential interactions. The Gibbs-Duhem equation for ternary systems (Eq. 23) was utilized to calculate the ΔG_2^{pref} values:

$$\Delta G_2^{pref} = \frac{m_1 \Delta G_1^{pref} + m_3 \Delta G_3^{pref}}{m_2} \tag{23}$$

$$\Delta G_2^{pref} = \mu_2^E(protein\ in\ water - organic\ mixtures) - \mu_2^E(pure\ protein) \tag{24}$$

where μ_1^E, μ_2^E, and μ_3^E are the excess chemical potentials of water, protein, and acetone, respectively; and m_1, m_2, and m_3 are the masses of water, protein, and acetone, respectively.

3.5. Effect of Preferential Interactions

The enzyme activity and sorption experiments can be summarized as follows:

1. Lysozyme is in the native (preferentially hydrated) state at high water content. Lysozyme has a higher affinity for water than for acetone. The ΔG_1^{pref}, ΔG_2^{pref}, and $(\partial g_3/\partial g_2)_{T,\mu1,\mu3}$ values are negative. On the other hand, the $(\partial g_1/\partial g_2)_{T,\mu1,\mu3}$, Z_1^E, and R_1^E values are positive. Enzyme activity with a high selectivity was observed in this range [11, 13].
2. The preferential binding of acetone to lysozyme was observed at intermediate water content. The $(\partial g_3/\partial g_2)_{T,\mu1,\mu3}$, ΔG_1^{pref}, and ΔG_2^{pref} values are positive. On the other hand, the $(\partial g_1/\partial g_2)_{T,\mu1,\mu3}$, Z_1^E, and R_1^E values are negative. Acetone-induced irreversible inactivation was found in this concentration range.
3. The deep dehydration of proteins leads to the formation of a rigid (glassy-like) state as a consequence of the proton-transfer phenomena and hydrogen bonding between ionizable and polar protein residues [1, 2, 6, 7, 55-68]. The potential of a solvent to establish hydrogen bonds has already been demonstrated to constitute a critical factor that controls the structure and state of dehydrated proteins at room temperature [45-48]. Thus, H-accepting and H-donating properties of acetone are anticipated to be essential for estimating the possible effects of organic molecules on the structure of the protein. When a disruption occurs with the hydrogen bond mediated protein-protein contact in the dehydrated protein, sorbate molecules (acetone or water) may discern the disrupted contact's H-donating and H-accepting fragments. Water (H-acceptor and H-donor), however, is capable of solvating both H-donating and H-accepting groups. An H-accepting acetone molecule is anticipated to prefer H-donating groups, whereas the remaining H-bond of the H-accepting partner will be solvated by water more efficiently. For this reason, it is anticipated that acetone molecules are ineffective in solvating the dehydrated protein alone. This results in the preferential hydration of lysozyme. Therefore, the $(\partial g_1/\partial g_2)_{T,\mu1,\mu3}$ values are positive

at low water content. Lysozyme retains high residual enzyme activity after incubation in water-poor acetone (Figures 3 and 5). Enzymes can catalyze the synthetic reactions (esterification) at low water content [11, 16-21].

CONCLUSION

We investigated preferential interactions of hen egg-white lysozyme with water-acetone mixtures. The stability of lysozyme depends significantly on the water content in acetone.

Three distinct schemes are operative in water-acetone mixtures:

1. Enzyme is preferentially hydrated at high water content
2. At intermediate water content, lysozyme has a higher affinity for acetone molecules than for water. Therefore, the interactions responsible for maintaining the native conformation of acetone are disrupted. Acetone augments the degree of irreversible inactivation of lysozyme in this concentration interval. The residual enzyme activity is close to zero in this region.
3. At low water content, the acetone molecules are preferentially excluded from the enzyme surface. This results in preferential hydration. Lysozyme demonstrates significant residual activity in water-poor acetone.

REFERENCES

[1] Gregory, R. B. (1995). Protein hydration and glass transition behavior. In: *Protein-Solvent Interactions* (Gregory, R. B., ed.), Marcel Dekker, New York, 191-264.

[2] Rupley, J. A., Careri, G. (1991). Protein hydration and function. *Adv. Protein Chem. 41*, 37-172.

[3] Kuntz, I. D., Kauzmann, W. (1974). Hydration of proteins and polypeptides. *Adv. Protein Chem*. 28, 239-345.

[4] Oleinikova, A., Smolin, N., Brovchenko, I., Geiger, A., Winter, R. (2005). Properties of spanning water networks at protein surfaces. *J. Phys. Chem. B. 109*, 1988-1998.

[5] Durchschlag, H., Zipper, P. (2001). Comparative investigations of biopolymer hydration by physicochemical and modeling techniques. *Biophys. Chem. 93*, 141-157.

[6] Sirotkin, V. A., Khadiullina, A. V. (2013). Gibbs energies, enthalpies, and entropies of water and lysozyme at the inner edge of excess hydration. *J. Chem. Phys. 139*, 075102/1-075102/9.

[7] Sirotkin, V. A., Khadiullina, A. V. (2014). A study of the hydration of ribonuclease A using densitometry: Effect of the protein hydrophobicity and polarity. *Chem. Phys. Lett, 603*, 13-17.

[8] Privalov, P. L., Crane- Robinson, C. (2017). Role of water in the formation of macromolecular structures. *Eur. Biophys. J. 46*, 203-224.

[9] Xu, M., Ermolenkov, V. V., Uversky, V. N., Lednev, I. K. (2008). Hen egg white lysozyme fibrillation: A deep-UV resonance Raman spectroscopic study. *J. Biophoton. 1*, 215-229.

[10] Rezaei-Ghaleh, N., Zwecktetter, M., Morshedi, D., Ebrahym-Habibi, A., Nemat-Gorgani, M. (2008). Amyloidogenic potential of α-chymotrypsin in different conformational states. *Biopolymers 91*, 28-36.

[11] Kijima, T., Yamamoto, S., Kise, H. (1996). Study of tryptophan fluorescence and catalytic activity of α-chymotrypsin in aqueous organic media. *Enz. Microb. Technol. 18*, 2-6.

[12] Sirotkin, V. A., Hüttl R., Wolf, G. (2005). Thermochemical investigations of hydrolysis of p-nitrophenyl acetate in water-acetonitrile mixtures, *Thermochim. Acta 426*, 1-6.

[13] Khmelnitsky, Yu., Mozhaev, V. V., Belova, A. B., Sergeeva, M. V., Martinek, K. (1991). Denaturation capacity: A new quantitative criterion for selection of organic solvents as reaction media in biocatalysis. *Eur. J. Biochem*. 198, 31-41.

[14] Partridge, J., Moore, B. D., Halling, P. J., (1999) α-Chymotrypsin stability in aqueous-acetonitrile mixtures: Is the native enzyme thermodynamically or kinetically stable under low water conditions? *J. Mol. Cat. B: Enzym. 6*, 11-20.

[15] Griebenow, K.; Klibanov, A. M. (1996). On protein denaturation in aqueous-organic mixtures but in pure organic solvents. *J. Amer. Chem. Soc. 118*, 11695-11700.

[16] Klibanov, A. M. (2001**).** Improving enzymes by using them in organic solvents, *Nature 409*, 241-246.

[17] Halling, P. J. (2004). What can we learn by studying enzymes in nonaqueous media? *Phil. Trans. R. Soc. Lond. B Biol. Sci. 359*, 1287– 1297.

[18] Lousa, D., Baptista, A. M., Soares, C. M. (2013). A molecular perspective on nonaqueous biocatalysis: Contributions from simulation studies. *Phys. Chem. Chem. Phys. 15*, 13723-13736.

[19] Carrea, G., Riva, S. (2000). Properties and synthetic applications of enzymes in organic solvents, *Angew. Chem. Int. Ed. 39*, 2226-2254.

[20] Micaelo, N. M., Soares, C. M. (2007). Modeling hydration mechanisms of enzymes in nonpolar and polar organic solvents, *FEBS J. 274*, 2424–2436.

[21] Clark, D. S. (2004). Characteristics of nearly dry enzymes in organic solvents: implications for biocatalysis in the absence of water, *Phil. Trans. R. Soc. Lond. B Biol. Sci. 359*, 1299–1307.

[22] Serdakowski, A. L., Dordick, J. S. (2007). Enzyme activation for organic solvents made easy, *Trends Biotechnol. 26*, 54-48.

[23] Rariy, R., Klibanov, A. M. (1997). Correct protein folding in glycerol, *Proc. Natl. Acad. Sci. U.S.A. 94*, 13520-13523.

[24] Timasheff, S. N. (2002). Protein-solvent preferential interactions, protein hydration, and the modulation of biochemical reactions by solvent components. *Proc. Natl. Acad. Sci. USA. 99*, 9721-9726.

[25] Arakawa, T., Kita, Y., Timasheff, S. N. (2007). Protein precipitation and denaturation by dimethyl sulfoxide. *Biophys. Chem. 131*, 62-70.

[26] Gekko, K., Ohmae, E., Kameyama, K., Takagi, T. (1998). Acetonitrile-protein interactions: Amino acid solubility and preferential solvation. *Biochim. Biophys. Acta 1387*, 195-205.

[27] Kovrigin, E. L., Potekhin, S. A. (2000). On stabilizing action of protein denaturants: Acetonitrile effect on stability of lysozyme in aqueous solutions. *Biophys. Chem. 83*, 45-59.

[28] Shimizu, S., Matubayasi, N. (2014). Preferential solvation: Dividing surface vs excess numbers. *J. Phys. Chem. B. 118*, 3922-3930.

[29] Auton, M., Bolen, D. W., Rosgen, J. (2008). Structural thermodynamics of protein preferential solvation: Osmolyte solvation of proteins, amino acids, and peptides. *Proteins 73*, 802–813.

[30] Casassa, E. F., Eisenberg, H. (1964). Thermodynamic analysis of multicomponent solutions. *Adv. Protein Chem. 19*, 287–395.

[31] Tanford, C. (1969.) Extension of theory of linked functions to incorporate effects of protein hydration. *J. Mol. Biol. 39*, 539– 544.

[32] Schellman, J. A. (1987). Selective binding and solvent denaturation. *Biopolymers 26*, 549–559.

[33] Arakawa, T., Bhat, R., Timasheff, S. N. (1990). Why preferential hydration does not always stabilize the native structure of globular proteins. *Biochem. 29*, 1924-1931.

[34] Smith, P. E. (2006). Equilibrium dialysis data and the relationships between preferential interaction parameters in biological systems in terms of Kirkwood-Buff integrals. *J. Phys. Chem. B 110*, 2862–2868.

[35] Parsegian, V. A., Rand, R. P., Rau, D. C. (2000). Osmotic stress, crowding, preferential hydration, and binding: A comparison of perspectives. *Proc. Natl. Acad. Sci. U.S.A. 97*, 3987–3992.

[36] Kirby Hade, E. P., Tanford, C. (1967). Isopiestic compositions as a measure of preferential interactions of macromolecules in two-component solvents. Application to proteins in concentrated aqueous cesium chloride and guanidine hydrochloride. *J. Amer. Chem. Soc. 89,* 5034-5040.

[37] Kamiyama, T., Liu, H. L., Kimura, T. (2009). Preferential solvation of lysozyme by dimethyl sulfoxide in binary solutions of water and dimethyl sulfoxide. *J. Therm. Anal. Cal. 95*, 353-359.

[38] Reisler, E., Haik, Y., Eisenberg, H. (1977). Bovine serum albumin in aqueous guanidine hydrochloride solutions. Preferential and absolute interactions and comparison with other systems. *Biochem. 16*, 197-203.

[39] Izumi, T., Yoshimura, Y., Inoue, H. (1980). Solvation of lysozyme in water/dioxane mixtures studied in the frozen state by NMR spectroscopy. *Arch. Biophys. Biochem. 200,* 444-451.

[40] Fersht, A. (1999). *Structure and Mechanism in Protein Science: A Guide to Enzyme Catalysis and Protein Folding;* Freeman & Co: New York.

[41] Lehninger, A. L.; Nelson, D. L.; Cox, M. M. (1993). *Principles of Biochemistry*; Worth: New York.

[42] Perrin, D. D., Armarego, W. L. F., Perrin, D. R. (1980). *Purification of Laboratory Chemicals*, Oxford: Pergamon Press.

[43] Borisover, M. D., Sirotkin, V. A., Solomonov, B. N. (1995). Isotherm of water sorption by human serum albumin in dioxane: Comparison with calorimetric data. *J. Phys. Org. Chem. 8,* 84-88.

[44] Sirotkin, V. A., Borisover, M. D., Solomonov, B. N. (1995). Heat effects and water sorption by human serum albumin on its suspension in water-dimethyl sulfoxide mixtures. *Thermochim. Acta, 256*, 175-183.

[45] Sirotkin, V. A., Kuchierskaya, A. A. (2017). Preferential solvation/hydration of α- chymotrypsin in water-acetonitrile mixtures. *J. Phys. Chem. B. 121*, 4422-4430.

[46] Sirotkin, V. A., Kuchierskaya, A. A. (2017). Lysozyme in water-acetonitrile mixtures: Preferential solvation at the inner edge of excess hydration. *J. Chem. Phys. 146*, 215101-8.

[47] Sirotkin, V. A., Kuchierskaya, A. A. (2017). α-Chymotrypsin in water-ethanol mixtures: Effect of preferential interactions. *Chem. Phys. Lett. 689*, 156-161.

[48] Sirotkin, V. A., Kuchierskaya, A. A. (2017). α-Chymotrypsin in water-acetone and water-dimethyl sulfoxide mixtures: Effect of preferential solvation and hydration. *Proteins: Functions, Structure and Bioinformatics, 85*, 1808-1819.

[49] Sirotkin V. A., Korolev D. V. (2005). Effect of acetonitrile on the hydration of human serum albumin films: a calorimetric and spectroscopic study. *Thermochim. Acta,* 432, 246-253.

[50] Atkins, P. W. (2006). *Physical Chemistry*. 8th ed. Oxford: Oxford University Press.

[51] Prausnitz, J. M. (1969). *Molecular Thermodynamics of Fluid-Phase Equilibria*. N. J.: Prentice-Hall, Inc., Engelwood Cliffs.

[52] Belousov, V. P., Panov, M. Y. (1994). *Thermodynamic properties of aqueous solutions of organic substances*. Boca Raton, Fla.: CRC Press.

[53] Kogan, V. B., Fridman, B. N., Kafarov, V. V. (1966). *Liquid–Vapor Equilibrium*. Moscow: Nauka.

[54] Bell, G., Janssen, A. E. M., Halling, P. (1996). Water activity fails to predict critical hydration level for enzyme activity in polar organic solvents: Interconversion of water concentrations and activities. *Enzym. Microb. Technol. 20*, 471-476.

[55] Bull, H. B. (1944). Adsorption of water vapor by proteins. *J. Amer. Chem. Soc*. 66, 1499-1507.

[56] Luscher-Mattli, M., Ruegg, M. (1982). Thermodynamic functions of biopolymer hydration. I. Their determination by vapor pressure studies, discussed in an analysis of the primary hydration process. *Biopolymers* 21, 403-418.

[57] Luscher-Mattli, M., Ruegg, M. (1982). Thermodynamic functions of biopolymer hydration. II. Enthalpy-entropy compensation in hydrophilic hydration process. *Biopolymers* 21, 419-429.

[58] Bone, S. (1987). Time-domain reflectometry studies of water binding and structural flexibility in chymotrypsin. *Biochem. Biophys. Acta.* 916, 128-134.

[59] Hnojewyj, W. S., Reyerson, L. H. (1961). Further studies on the sorption of H_2O and D_2O vapors by lysozyme and the deuterium-hydrogen exchange effect. *J. Phys. Chem.* 65, 1694-1698.

[60] Sirotkin, V. A., Khadiullina, A. V. (2011). Hydration of proteins: excess partial enthalpies of water and proteins. *J. Phys. Chem. B*, 115, 15110-15118.

[61] Sirotkin, V. A., Komissarov, I. A., Khadiullina, A. V., (2012). Hydration of proteins: excess partial volumes of water and proteins, *J. Phys. Chem. B*, 116, 4098-4105.

[62] Sirotkin, V. A. (2005). Effect of dioxane on the structure and hydration-dehydration of α-chymotrypsin as measured by FTIR spectroscopy. *Biochim. Biophys. Acta.*, 1750, 17-29.

[63] Kocherbitov, V., Arnebrant, T. (2010). Hydration of Lysozyme: The Protein-Protein Interface and the Enthalpy-Entropy Compensation. *Langmuir,* 26, 3918-3922.

[64] Foss, J. G., Reyerson, L. H. (1958). The sorption of water vapor by lyophilized ribonuclease. *J. Phys. Chem.* 62, 1214-1216.

[65] Amberg, C. H. (1957) Heats of adsorption of water vapor on bovine serum albumin. *J. Amer. Chem. Soc,* 79, 3980-3984.

[66] Almog, R., Schrier, E. E. (1978). The dependence of the heat of solution of lyophilized solid ribonuclease A on water content. *J. Phys. Chem,* 82, 1701-1702.

[67] Smith, A. L., Shirazi, H. M., Mulligan, S. R. (2002). Water sorption isotherms and enthalpies by lysozyme using the quartz crystal microbalance/heat conduction calorimeter. *Biochim. Biophys. Acta,* 1594, 150-159.

[68] Kocherbitov, V., Arnebrant, T., Söderman, O. (2004). Lysozyme-water interactions studied by sorption calorimetry. *J. Phys. Chem. B,* 108, 19036-19042.

In: Acetone
Editors: Irenka Kozłowska et al.
ISBN: 978-1-53613-298-4

Chapter 3

INTERACTION ENTHALPIES OF α-CHYMOTRYPSIN WITH WATER-ACETONE MIXTURES

Vladimir A. Sirotkin[*] and Alexandra A. Kuchierskaya
A. M. Butlerov Institute of Chemistry,
Kazan Federal University, Kazan, Russia

ABSTRACT

The aim of our investigation is to provide a thermochemical description of the stabilizing/destabilizing effects of organic solvents on the enzyme stability to elucidate what intermolecular processes produce the main effect on the stability and functions of the enzymes at low, intermediate, and high water content in organic liquids. Bovine pancreatic α-chymotrypsin was utilized as a model enzyme. Isothermal calorimetry measurements were performed to study the stability of α-chymotrypsin in water-acetone mixtures.

Interaction enthalpies of α-chymotrypsin with water-acetone mixtures have been measured at 25^{0}C. Our results clearly show that isothermal calorimetry is an effective experimental tool for studying the

[*] Corresponding Author Email: vsir@mail.ru.

simultaneous action of water and organic liquids on the stability of enzyme macromolecules. Thermochemical results were compared with the enzyme activity data. The degree of stabilization/destabilization of α-chymotrypsin depends significantly on the water content in acetone. There are three concentration regimes. At high water content, the residual enzyme activity values are close to 100%. At intermediate water content, residual enzyme activity is close to zero. The residual catalytic activity in the water-poor acetone is ~80%.

Keywords: enzyme hydration, calorimetry, enthalpy, water, organic solvent, acetone, preferential solvation

1. Introduction

Enzyme-water interactions play an essential role in determining the functions and stability of enzyme macromolecules [1-8]. Knowledge of processes that occur upon protein hydration in the presence of organic solvents is critical in numerous biomedical and biotechnological applications [9-21]. For example, there are numerous benefits in employing the enzymes in the water-poor organic liquids [14-21].

i) The solubility of hydrophobic substrates is higher.
ii) The undesirable side reactions caused by water are suppressed.
iii) Hydrolytic enzymes in nonaqueous media can catalyze the reversed hydrolytic reactions, such as transesterification, esterification, and peptide synthesis.

Therefore, elucidating the precise mechanism of protein hydration in the presence of organic solvents necessitates effective experimental methods that reveal thermodynamic information regarding enzyme–organic and enzyme–water interactions.

Water sorption isotherms (water content as a function of water activity) of numerous proteins have been reliably measured [1-3, 22-33]. Typical water sorption isotherm has a sigmoid shape. Foss and Reyerson studied the water sorption by ribonuclease in the range of 0-8% (g g^{-1}) [34]. From the temperature dependence of the water sorption isotherms in the range 20-30°C, the differential entropies of water sorption were estimated. Luscher-Mattli and Ruegg [23, 24] calculated the entropies of water sorption by lysozyme and α-chymotrypsin. They investigated the water sorption isotherms in the water mass fraction range from 0 to 0.4. The hydration entropies were calculated from the temperature dependence of the water vapor pressure in the range of 10-40°C. Bone studied the water sorption by lysozyme in the range of 1.5–19% (g g^{-1}) [25] Calculations were performed using the temperature dependence of the water vapor pressure in the range of 6–46°C.

Isothermal calorimetry is one of the most effective methods for obtaining reliable thermochemical information on the protein-water interaction in various environments. Amberg measured the heats of water vapor sorption using bovine serum albumin within a water content range of 0-12% (g g^{-1}) at 20°C [36]. The solution enthalpies of the dried and hydrated ribonuclease A (RNase A) were measured using isothermal calorimetry at 25°C [36]. The water content of RNase A was varied from 0 to 26% (g g^{-1}), and the solution enthalpy was negative (exothermic). As the water content of ribonuclease A augments, the solution enthalpy approaches zero. Smith et al., measured the heat effects of water sorption by lysozyme in the range of relative water vapor pressures from 0 to 0.895 [37]. He obtained both the sorption isotherm and enthalpy of hydration of the enzyme within a water content range of 0-18% (g g^{-1}) at 25°C. Sorption calorimetry has been utilized to measure the water sorption isotherm for lysozyme and the corresponding heat effects in the entire range of water activity [38].

The most important observations of these calorimetric studies can be summarized as follows:

1) The hydration enthalpies vary strongly with the hydration level.
2) The hydration enthalpies are highly exothermic at low hydration level.

3) As the water content increases, the hydration enthalpies approach the enthalpy of condensation of pure water.

Thermochemical studies may be extremely pertinent to the understanding of the enzyme-water interactions in organic liquids. The interaction enthalpies of α-chymotrypsin with neat organic solvents (hydrocarbons, alcohols, and hydrogen bond accepting solvents) have been studied using isothermal calorimetry [39]. It was found that the interaction enthalpies depend cooperatively on the solvent hydrophilicity. All the solvents were divided into two groups. The first group included the hydrophilic solvents (methanol, DMSO). The interaction enthalpies of the dried enzyme with hydrophilic liquids were strongly exothermic. The second group included the hydrophobic and medium hydrophilic liquids (benzene, dioxane, butanol-1, propanol-1). The interaction enthalpies for the second group of solvents were close to zero.

The main focus of our investigation is to estimate the organic solvent effect on the enzyme hydration in the entire range of thermodynamic water activity by isothermal calorimetry. The aim of our work was to develop an experimental approach for separating the enzyme-water and enzyme-organic solvent contributions to the interaction ethnalpies of enzyme macromolecules in aqueous-organic mixtures.

Bovine pancreatic α-chymotrypsin (CT) was utilized as a model enzyme. It is one the most studied and applied in biophysics and enzymology [40, 41]. Physiological role of CT is to hydrolyze peptide bonds [40, 41]. α-Chymotrypsin is an example of a predominantly β-sheet protein.

The choice of acetone was determined for the following reasons:

1) Acetone is an organic solvent that is water-miscible. Consequently, the effect of this substance on the hydration and structure of α-chymotrypsin in the entire range of water content can be investigated.

2) Acetone is able to form hydrogen bonds with several hydrogen donors. However, opposite to water, it does not possess hydrogen bond donating ability.

2. EXPERIMENTAL

2.1. Materials

Bovine pancreatic α-chymotrypsin was purchased from Sigma Chemical Co. (St. Louis, MO, U.S.A., No. C 4129, protein content >95%; essentially salt free; EC 3.4.21.1; 66 units of specific activity for N-benzoyl-L-tyrosine ethyl ester). The molecular mass of bovine pancreatic α-chymotrypsin was determined to be 25000 Da. Acetone (analytical grade, with a purity of >99%) was purified and dried according to the recommended guidelines [42]. Water used was doubly distilled. All water-organic mixtures were prepared gravimetrically using a Precisa balance (Swiss) with a precision of 0.00001 g.

2.2. Thermodynamic Activity of Water in Organic Solvent

Water activity (a_W) in organic solvent was calculated using Eq. 1:

$$a_w = \gamma_w x_w \tag{1}$$

where x_W is the mole fraction of water in the solution and γ_W is the activity coefficient of water (in mole fraction; the standard state is pure water).

Water activity coefficients (γ_W) were calculated using literature data [43, 44] on the vapor-liquid equilibrium according to Eq. 2.

$$\gamma_w = \frac{y_w P_{tot}}{x_w P_w^o} \tag{2}$$

where y_W is the measured mole fraction of water in vapor phase, P_{tot} is the total pressure, P_o is the saturated vapor pressure of pure water at the same temperature and x_W is the mole fraction of water in the liquid phase.

2.3. Water Sorption Measurements

The dried protein preparation *(zero hydration level)* was obtained through vacuum drying using a microthermoanalyzer "Setaram" MGDTD-17S (±0.00001 g) at 25°C and 0.1 Pa, until a constant sample mass was reached. The water content of the dried protein was estimated as 0.002 ± 0.001 g water g^{-1} protein using the Karl Fischer titration method, according to the recommended guidelines [29, 30, 42].

The protein preparations were prepared as follows. The initially dehydrated protein samples were exposed to pure water vapor. The water vapor was consecutively flowed through a thermostated glass tube with drying agent (P_2O_5), a thermostated saturator filled with saturated salt solution, and a cell containing the protein sample. Schematic representation of the experimental set-up is given in [32]. Sorption equilibrium was reached after 6 h. The water activity (a_w) in the vapor phase was adjusted by changing the saturated salt solution in the saturator. The water activities of the saturated salt and KOH solutions were taken from [45, 46]. The following substances were used: LiBr (a_w = 0.064), KOH (a_w = 0.078), LiCl (a_w = 0.11), $CaBr_2$ (a_w = 0.17), CH_3COOK (a_w = 0.22), $MgCl_2$ (a_w = 0.33), K_2CO_3 (a_w = 0.44), $Mg(NO_3)_2$ (a_w = 0.53), NaCl (a_w = 0.75), KCl (a_w = 0.84), KNO_3 (a_w = 0.94), and K_2SO_4 (a_w = 0.97). The salts used for sample conditioning were of analytical pure grade. The water content of the samples after equilibration was measured by drying under vacuum using a microthermoanalyzer "Setaram" MGDTD-17S at 25°C and 0.1 Pa, until a constant sample mass was reached.

2.4. Isothermal Calorimetry

Heat effects were measured on a BT-2.15 "Setaram" calorimeter at 25°C. The calorimeter was calibrated using the Joule effect and tested by measuring the solution enthalpy of potassium chloride in water. Water-organic mixtures were prepared from aqueous buffer (10 mM phosphate at pH 7.0) by weighing.

The interaction enthalpies of α-chymotrypsin with aqueous–organic mixtures were measured as follows. A protein sample (5–10 mg) was first loaded into a titanium container with Teflon gaskets. After sealing, the container was placed into the calorimetric cell preliminarily filled with a solvent (4.0 ml). When thermal equilibrium was attained, Teflon gaskets were pierced, and the solvent began to interact with the protein. The procedures for heat effect measurements are described in detail in [39].

The dried protein preparation *(zero hydration level)* was obtained through vacuum drying using a microthermoanalyzer "Setaram" MGDTD-17S at 25°C and 0.1 Pa, until a constant sample mass was reached.

2.5. Enzyme Activity

Residual enzyme activity was determined by measuring the enzyme activity after storage in water-organic mixtures as described previously [47, 48]. The model process was the hydrolysis of N-acetyl-L-tyrosine ethyl ester (ATEE) catalyzed by α-chymotrypsin. The measurements were performed on a TitroLine Alpha Plus potentiometric titrator (Schott Instruments, Germany) in the pH-static mode at pH 8.0 and 25°C. The concentration of the substrate was 4.0×10^{-3} mol/L. In the course of the experiment, the pH value was maintained at a constant level by adding a titrant (a potassium hydroxide solution of known concentration), which neutralized the acid (N-acetyl-L-tyrosine) released during the hydrolysis. The kinetic curve obtained was the time dependence of the amount of the reagent spent for titrating the acid released. Each kinetic curve was reproduced not less than three times.

The reaction mixture was prepared as follows. The α-chymotrypsin sample was immersed in an aqueous-organic mixture of required composition, and was incubated at 25°C for 180 min. The concentration of the α-chymotrypsin in the water-organic mixture was 1 mg/ml. Adding 100-μl aliquots of the α-chymotrypsin solution in the water-rich acetone (or the α-chymotrypsin suspension in the water-poor acetone) to the aqueous solution of the substrate, we initiated the enzymatic reaction.

3. Results and Discussion

3.1. Water Sorption Isotherm

Figure 1 shows the water sorption isotherm for α-chymotrypsin in the absence of organic solvent at 25°C. The initial state was the protein dried in a flow of air at a water activity of 0.01 and 25°C.

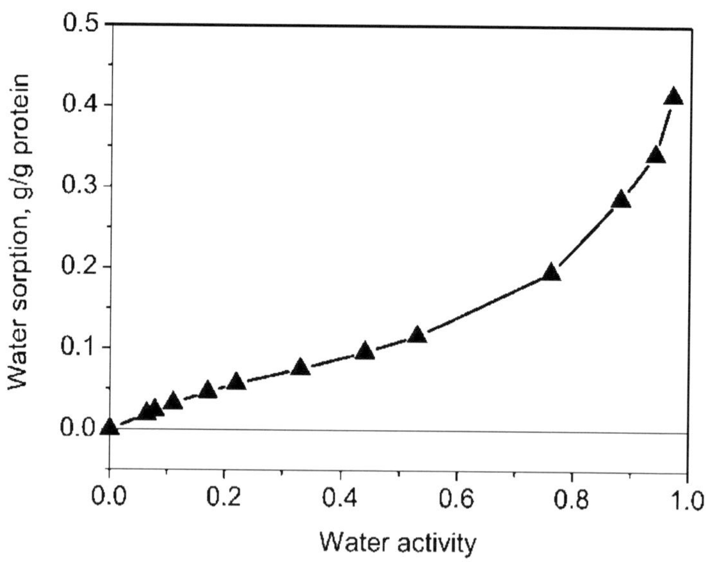

Figure 1. α-Chymotrypsin. Water sorption isotherm at 25°C.

As concluded from Figure 1, the water sorption isotherm resembles a typical sigmoid shaped curve. This result is in accordance with those for α-chymotrypsin and other proteins [1, 2, 22-27].

3.2. Interaction Enthalpies of α-Chymotrypsin with Water in the Absence of Organic Solvent

The heat effects (ΔH_{tot}) of the α-chymotrypsin interaction with water in the absence of organic solvent were determined as follows. The initial state was the protein dried in air at water activity less than 0.01 and 25°C. The final state was the protein at equilibrium with pure water vapor with the varied water activity at 25°C and atmospheric pressure.

Thermodynamic data of two types were used to determine the ΔH_{tot} values in the absence of organic solvent.

3.2.1. Isothermal Calorimetry Data

The ΔH_{tot} values were determined in this work using isothermal calorimetry (Eq. 3-5). The studied systems can be described as follows:

Initial state for the measured enthalpy changes, ΔH_{tot}:

1) **Enzyme:**
   i) The enzyme obtained by drying in the absence of organic solvent at water activity less than 0.01 at 25°C ($\overline{H}_E(dried)$);
   ii) The enzyme hydrated by pure water vapor with the varied water activity in the absence of organic solvent at 25°C ($\overline{H}_E(hydrated)$).
2) **Water:** The initial state was the water molecules bound to enzyme at 25°C ($\overline{H}_w(enzyme)$). Final state for the measured enthalpy changes, ΔH_{tot}:
3) **Enzyme:** The final state was the enzyme dissolved in liquid water at 25°C and atmospheric pressure ($\overline{H}_E(solvent)$).

4) **Water:** The final state was the water molecules dissolved in liquid water at 25°C and atmospheric pressure ($\overline{H}_w(solvent)$). Symbols used in Section 3.2 are described in Table 1.

Table 1. Symbols used in Section 3.2

Symbol	**Name**	**Unit**
$\overline{H}_E(solvent)$	Partial enthalpy of water in the liquid at 25°C	J/mol
$\overline{H}_w(enzyme)$	Partial enthalpy of water bound to the enzyme at 25°C	
$\overline{H}_w(gas)$	Partial enthalpy of water in gas phase at 25°C	
$\overline{H}_E(dried)$	Partial enthalpy of the dried enzyme at 25°C	J/mol enzyme
$\overline{H}_E(hydrated)$	Partial enthalpy of the hydrated enzyme at 25°C	
$\Delta H_{tot}(dried)$	Interaction enthalpy of the initially dried enzyme with pure liquid water	J
$\Delta H_{tot}(hydrated)$	Interaction enthalpy of the initially hydrated enzyme with pure liquid water	
$\Delta H_{vap}^{H_2O}$	Enthalpy of vaporization of water at atmospheric pressure and 25°C	kJ/mol water
$\Delta H_{sorp}^{H_2O/gas}$	Enthalpy of transfer of water from the gas phase to the enzyme	

The interaction enthalpy of the dried enzyme with pure liquid water is described by Eq. 3:

$$\Delta H_{tot}(dried) = n_E \overline{H}_E(solvent) - n_E \overline{H}_E(dried) \tag{3}$$

The interaction enthalpy of the hydrated enzyme with pure liquid water is given by Eq. 4:

$$\Delta H_{tot}(hydrated) = n_w \overline{H}_w(solvent) + n_E \overline{H}_E(solvent) - n_E \overline{H}_E(hydrated) - n_w \overline{H}_w(enzyme) \tag{4}$$

where n_E and n_w are the moles of the enzyme and water.

The $\Delta H_{tot}(dried)$ values demonstrate the enzyme–water (bulk solvent) interactions. The $\Delta H_{tot}(hydrated)$ values describe simultaneously the enzyme-water (bulk solvent) and the enzyme–sorbed water interactions. The $\Delta H_{tot}(dried) - \Delta H_{tot}(hydrated)$ value includes the contributions resulting from (i) the difference between the partial enthalpies of the dried and hydrated enzyme and (ii) the difference between the partial enthalpies of water bound to enzyme and water in the bulk solvent.

To characterize the enzyme–sorbed water interactions, the $\Delta H_{tot}(dried)$ and $\Delta H_{tot}(hydrated)$ values should be compared with the water sorption data obtained independently, i.e., n_w.(Figure 1).

$$\Delta H_{tot}(dried) - \Delta H_{tot}(hydrated) = n_E(\bar{H}_E(hydrated) - \bar{H}_E(dried)) + n_w(\bar{H}_w(gas) - \bar{H}_w(solvent)) + n_w(\bar{H}_w(enzyme) - \bar{H}_w(gas)) = n_w(\Delta H_{sorp}^{H_2O/gas} + \Delta H_{vap}^{H_2O}) \quad (5)$$

where $\Delta H_{sorp}^{H_2O/gas}$ is the water sorption enthalpy (the transfer enthalpy of water from the gas phase to the enzyme); $\Delta H_{vap}^{H_2O}$ is the enthalpy of water vaporization at 25°C and atmospheric pressure (43.7 kJ/mol water [45]); $\bar{H}_w(gas)$ is the partial enthalpy of water in gas phase at 25°C and atmospheric pressure.

The $\Delta H_{sorp}^{H_2O/gas}$ value (Reference state is pure water in the gas phase at 25°C and atmospheric pressure) is calculated using Eq. 6:

$$\Delta H_{sorp}^{H_2O/gas} = \frac{\Delta H_{tot}(dried) - \Delta H_{tot}(hydrated)}{n_w} - \Delta H_{vap}^{H_2O} = \frac{n_E}{n_w}(\bar{H}_E(hydrated) - \bar{H}_E(dried)) + (\bar{H}_w(enzyme) - \bar{H}_w(gas)) \quad (6)$$

The $\Delta H_{sorp}^{H_2O/gas}$ value is called sometimes the enthalpy of protein hydration (or water sorption enthalpy). The $\Delta H_{sorp}^{H_2O/gas}$ values determined in this work are given in Figure 2.

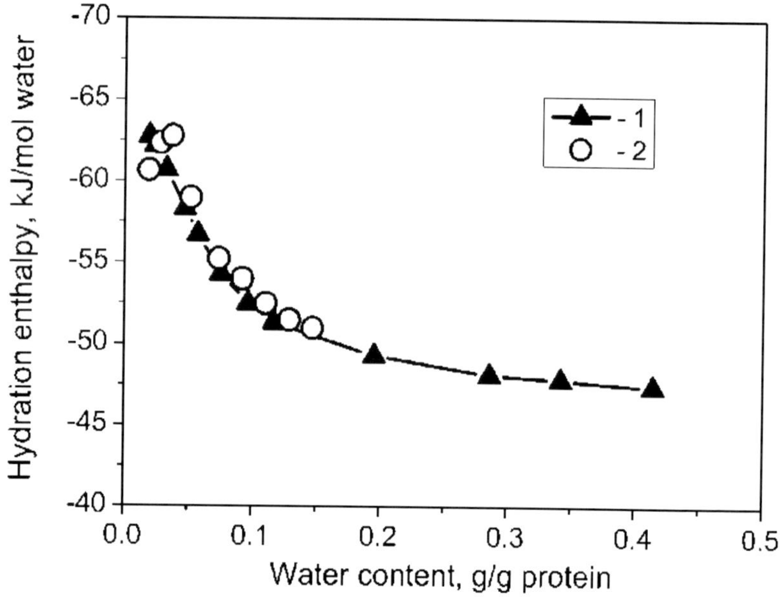

Figure 2. (1) α-Chymotrypsin. The $\Delta H_{sorp}^{H_2O/gas}$ values as a function of the enzyme hydration at 25°C. This work. (2) Chymotrypsinogen. The $\Delta H_{sorp}^{H_2O/gas}$ values estimated from isosteric data [49].

The hydration enthalpy (Reference state is pure liquid water at 25 °C) can be estimated as follows (Eq. 7):

$$\Delta H_{sorp}^{H_2O/H_2O\ (liquid)} = \frac{\Delta H_{tot}(dried) - \Delta H_{tot}(hydrated)}{n_w} =$$

$$\frac{n_E}{n_w}(\overline{H}_E(hydrated) - \overline{H}_E(dried)) + (\overline{H}_w(enzyme) - \overline{H}_w(gas)) \quad (7)$$

As concluded from Eqs. 6 and 7, this method does not separate the protein and water enthalpic contributions to the calorimetrically measured heat effects. According to Eq. 7, the $\Delta H_{sorp}^{H_2O/gas}$ value contains two components. The first component, $\frac{n_E}{n_w}(\overline{H}_E(hydrated) - \overline{H}_E(dried))$, is due to the changes in the protein state during hydration (including conformational rearrangements and glass transition). The second component, $\overline{H}_w(enzyme) - \overline{H}_w(gas)$, is due to the difference between the water state in the gas phase and the bound state.

3.2.2. Isosteric Quantities

The isosteric quantity is the change in partial molar quantity for the transfer of water from the vapor phase to the protein at water equilibrium pressure **P**. These values are estimated using Eq. 8. Eq. 8 is applied for constant (isosteric) water content (Γ).

$$lnP_\Gamma = \frac{\Delta\bar{S}_\Gamma}{R} - \frac{\Delta\overline{H}_\Gamma}{R}\left(\frac{1}{T}\right) = \frac{\bar{S}_w(gas)-\bar{S}_w(enzyme)}{R} - \frac{\overline{H}_w(gas)-\overline{H}_w(enzyme)}{R}\left(\frac{1}{T}\right) \quad (8)$$

where $\bar{S}_w(gas)$ is the partial molar entropy of water in the vapor phase; $\bar{S}_w(enzyme)$ is the partial molar entropy of the sorbed water; $\overline{H}_w(gas)$ is the partial molar enthalpy of water in the vapor phase; $\overline{H}_w(enzyme)$ is the partial molar enthalpy of the sorbed water; and R is the gas constant.

The isosteric quantities are estimated from the temperature dependence of the water sorption isotherm (The van't Hoff plots [lnP versus 1/T] at various protein hydration values.

To show the reliability of our experiments and calculations, the isosteric enthalpies of sorption of pure water on chymotrypsinogen ($\Delta\overline{H}_\Gamma$) from Ref. [49] are presented in Figure 2.

3.3. Interaction Enthalpies of α-Chymotrypsin with Water-Acetone Mixtures

The $\Delta H_{tot}(W)$ values (pure water sorption, Figure 3, data set no. 1) were estimated using Eq. 9 (Reference state is pure liquid water at 25°C):

$$\Delta H_{tot}(W) = (\Delta H_{tot}(dried) - \Delta H_{tot}(hydrated)\ )/n_E \qquad (9)$$

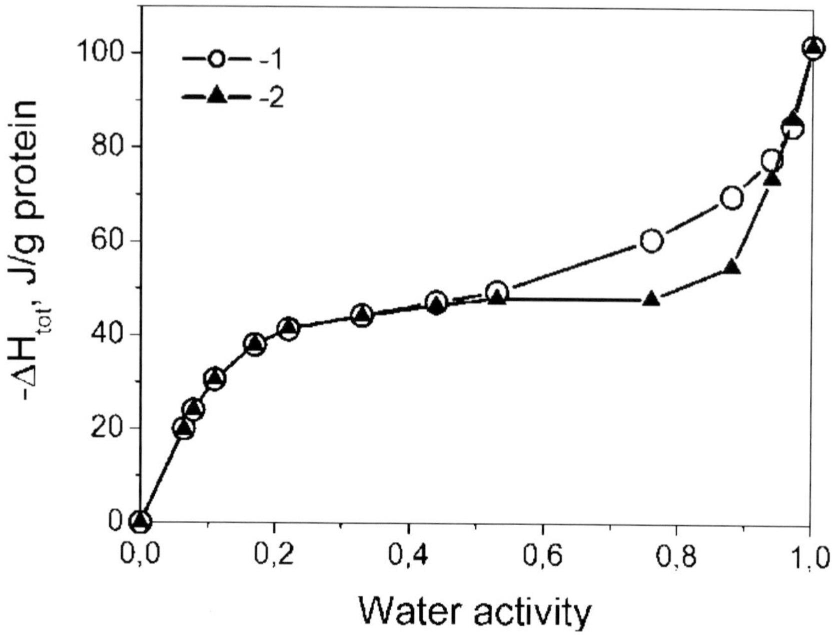

Figure 3. Interaction enthalpy of α-chymotrypsin with water (ΔH_{tot}) in the absence (*1*) and presence (2) of organic solvent at 25°C.

The heat effects ($\Delta H_{tot}(S)$, J/mol protein) of the α-chymotrypsin interaction with water-acetone mixtures were calculated follows. The initial state was the protein dried in a flow of air at a water activity of 0.01 and 25°C. The final state was the protein in an aqueous–organic mixture with a variable water activity at 25°C.

The dependences of the heat effects of interaction of α-chymotrypsin with pure water and water–acetone mixtures versus water activity are shown in Figure 3. At low water activity, the ΔH_{tot} values are close to zero. At intermediate and high water activity, the ΔH_{tot} values differ essentially from zero.

The effect of organic solvent on the interaction enthalpies of water with α-chymotrypsin was characterised by the difference in the ΔH_{tot} values obtained in the presence and absence of organic solvent at a given

a_w. The ΔH_{tot} function obtained in the presence of organic solvent has been compared with the ΔH_{tot} values determined for α-chymotrypsin in the absence of organic solvent. This comparison can show any effects of the organic solvent molecules on the interaction enthalpies of α-chymotrypsin with water. The difference between the ΔH_{tot} values obtained in the presence and absence of organic solvent ($\Delta\Delta H_{tot} = \Delta H_{tot}(S) - \Delta H_{tot}(W)$) is presented in Figure 4.

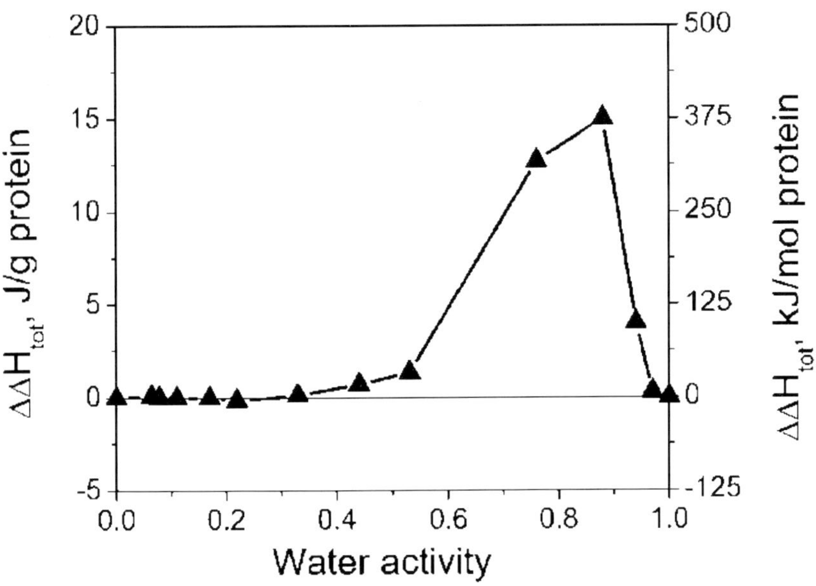

Figure 4. Difference interaction enthalpy ($\Delta\Delta H_{tot}$) as a function of water activity (a_w).

There are three composition regimes in Figure 4:

1) At water activity below 0.4, organic solvent does not influence the thermochemical characteristics of α-chymotrypsin.
2) At intermediate water activity, the difference thermochemical values differ essentially from zero.

3) At high water activity ($a_w > 0.9$), the difference thermochemical values are close to zero.

3.4. Residual Enzyme Activity

Figure 5 demonstrates typical kinetic curves for the enzymatic reaction catalyzed by α-chymotrypsin preliminarily incubated in water-acetone mixtures. The catalytic activity was characterized by the ratio of the extent of hydrolysis attained within 200 s with α-chymotrypsin incubated in a water-organic mixture to the same quantity measured using CT incubated in pure water (Figure 5, curve 1).

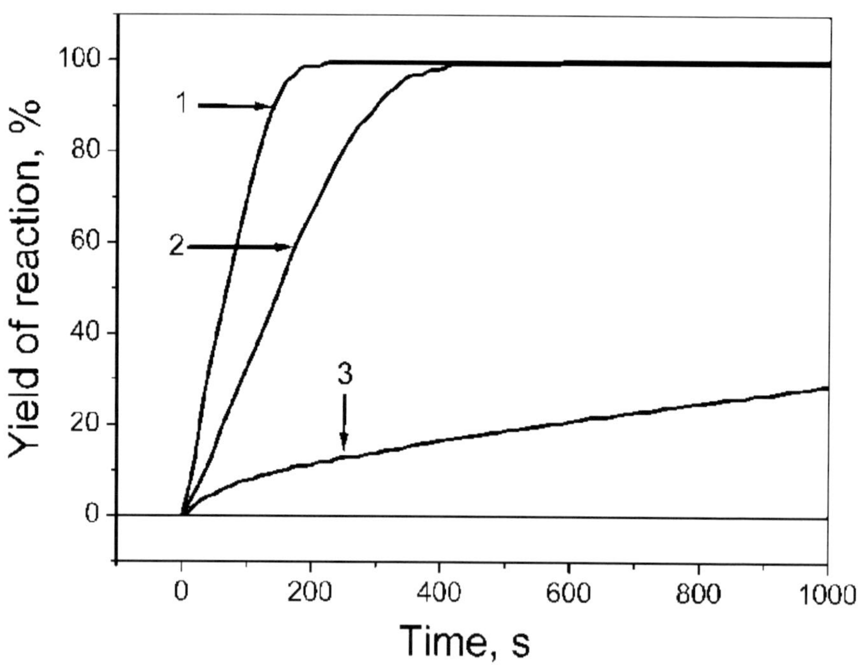

Figure 5. Typical kinetic curves for the hydrolysis of N-acetyl-L-tyrosine ethyl ester catalyzed by α-chymotrypsin preliminary incubated in water-acetone mixtures. Water activity: (1) 1.0, (2) 0.01, (3) 0.87.

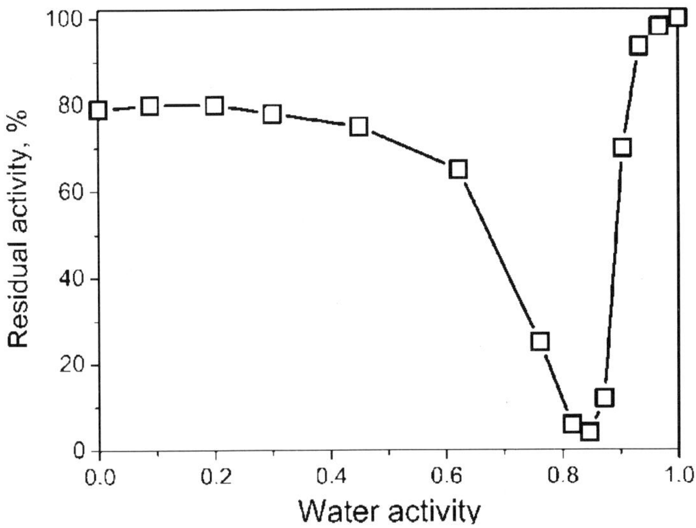

Figure 6. Residual enzyme activity (Hydrolysis of N-acetyl-L-tyrosine ethyl ester by α-chymotrypsin preliminary incubated in water-acetone mixtures).

The residual activity values are shown in Figure 6. As shown in Figure 6, acetone affects the residual enzyme activity of α-chymotrypsin in a complicated manner. The residual activity values are close to 100% at high water content (a_w~0.9-1.0). There is a sharp transition from the water-rich range to the intermediate one at $a_w < 0.9$. The residual enzyme activity of α-chymotrypsin changes from 100 to 0% in the transition region (a_w ~0.8-0.9). A minimum on the residual activity curve was found at $a_w \sim 0.8$. The residual activity augments at $a_w < 0.6$. At low water content, the residual enzyme activity is ~80%, compared with that observed after incubation in pure water.

3.5. Effect of Acetone on the Thermochemical Properties of α-Chymotrypsin

3.5.1. Regime 1 ($a_w = 0 - 0.4$)

At low water content, the dried proteins are in a glassy-like (rigid) state [1, 2, 51, 52]. The potential of a solvent to form hydrogen bonds is an important factor that controls the stability and structure of the dehydrated proteins at room temperature [39, 51, 52]. Thus, H-donating and H-accepting properties of organic solvents are expected to be essential for estimating the effects of organic molecules on the dehydrated protein. Acetone may be considered as an informative tool for analyzing the effect of the hydrogen bond accepting ability on the dehydrated protein stability.

Acetone is a moderate-strength hydrogen bond accepting solvent. No significant heat effects and structural rearrangements were observed for dioxane and acetonitrile, the moderate-strength hydrogen bond accepting solvents [39]. This effect can be explained as follows. When a hydrogen bond mediated protein-protein contact in the dehydrated protein is disrupted, sorbate molecules (organic solvent or water) may differentiate between H-donating and H-accepting fragments of the disrupted contact. Water (H-donor and H-acceptor) is able to solvate both H-accepting and H-donating groups. An H-accepting acetone molecule is expected to prefer H-donating groups, while the remaining H-bond of the H-accepting partner will be solvated by water more effectively. Hence, it is expected that acetone molecules are not effective in solvating the dehydrated protein alone. This indicates that the acetone molecules (similar to dioxane and acetonitrile [32, 39, 51] are preferentially excluded from the enzyme surface, producing preferential hydration. For this reason, no noticeable difference heat effects were observed when chymotrypsin was introduced into water-free acetone (Figure 3).

It follows that the main process that occurs at water activity from 0 to 0.4 is the preferential hydration of the dried protein. According to the suggested model, the sorption process then occurs as the penetration of water molecules into the structure of the dried (glassy-like) protein, the rupture of interprotein contacts, the hydration of the polar groups of these contacts, and the formation of new sorption sites.

3.5.2. Regime 2 (a_w = 0.4-0.9)

The difference thermochemical characteristics are sensitive to changes in the mechanism of water sorption (Figure 4). Noticeable differences in the mechanism of the enzyme hydration were observed in the presence of organic solvent. At water activity of 0.4-0.9, the rupture and hydration of the polar fragments of the interprotein contacts are followed by the increased endothermic heat effects. Note that each acetone molecule contains two methyl hydrophobic groups. We can therefore assume that, when inserted into the structure of the hydrated protein sorbent, acetone (unlike water) can solvate polar groups not only in hydrophilic but also in hydrophobic protein regions with the formation of new sorption sites for water.

To summarize, according to the proposed model, the special feature of the enzyme hydration at intermediate water activitiy is the penetration of the acetone molecules predominantly into hydrophobic protein regions, the rupture of the dehydration-induced protein-protein contacts, and the creation of new sorption sites. A maximum on the difference thermochemical curve (Figure 4) is the evidence in favor of this sorption mechanism.

These facts indicate that the preferential solvation of α-chymotrypsin is a dominating process in this water activity region. Acetone shows a destabilizing effect at intermediate water content. Acetone augments the degree of irreversible inactivation of α-chymotrypsin in this region (Figure 6) [48].

3.5.3. Regime 3 (a_w = 0.9-1.0)

α-Chymotrypsin is in the native (preferentially hydrated) state at high water content in acetone [48]. Acetone demonstrates a stabilizing effect on the residual activity at high water content. The residual activity values are close to 100% (Figure 6) [48]. The difference thermochemical values are close to zero (Figure 4).

REFERENCES

[1] Gregory, R. B. (1995). Protein hydration and glass transition behavior. In: *Protein-Solvent Interactions* (Gregory, R. B., ed.), Marcel Dekker, New York, 191-264.

[2] Rupley, J. A., Careri, G. (1991). Protein hydration and function. *Adv. Protein Chem. 41*, 37-172.

[3] Kuntz, I. D., Kauzmann, W. (1974). Hydration of proteins and polypeptides. *Adv. Protein Chem. 28*, 239-345.

[4] Oleinikova, A., Smolin, N., Brovchenko, I., Geiger, A., Winter, R. (2005). Properties of spanning water networks at protein surfaces. *J. Phys. Chem. B. 109*, 1988-1998.

[5] Durchschlag, H., Zipper, P. (2001). Comparative investigations of biopolymer hydration by physicochemical and modeling techniques. *Biophys. Chem. 93*, 141-157.

[6] Sirotkin, V. A., Khadiullina, A. V. (2013). Gibbs energies, enthalpies, and entropies of water and lysozyme at the inner edge of excess hydration. *J. Chem. Phys. 139*, 075102/1-075102/9.

[7] Sirotkin, V. A., Khadiullina, A. V. (2014). A study of the hydration of ribonuclease A using densitometry: Effect of the protein hydrophobicity and polarity. *Chem. Phys. Lett*, *603*, 13-17.

[8] Privalov, P. L., Crane‐Robinson, C. (2017). Role of water in the formation of macromolecular structures. *Eur. Biophys. J. 46*, 203-224.

[9] Kijima, T., Yamamoto, S., Kise, H. (1996). Study of tryptophan fluorescence and catalytic activity of α-chymotrypsin in aqueous organic media. *Enz. Microb. Technol. 18*, 2-6.

[10] Sirotkin, V. A., Hüttl, R., Wolf, G. (2005). Thermochemical investigations of hydrolysis of p-nitrophenyl acetate in water-acetonitrile mixtures, *Thermochim. Acta 426*, 1-6.

[11] Khmelnitsky, Yu., Mozhaev, V. V., Belova, A. B., Sergeeva, M. V., Martinek, K. (1991). Denaturation capacity: A new quantitative criterion for selection of organic solvents as reaction media in biocatalysis. *Eur. J. Biochem. 198*, 31-41.

[12] Partridge, J., Moore, B. D., Halling, P. J. (1999). α-Chymotrypsin stability in aqueous-acetonitrile mixtures: Is the native enzyme thermodynamically or kinetically stable under low water conditions? *J. Mol. Cat. B: Enzym. 6*, 11-20.

[13] Griebenow, K.; Klibanov, A. M. (1996). On protein denaturation in aqueous-organic mixtures but in pure organic solvents. *J. Amer. Chem. Soc. 118*, 11695-11700.

[14] Klibanov, A. M. (2001). Improving enzymes by using them in organic solvents, *Nature 409*, 241-246.

[15] Halling, P. J. (2004). What can we learn by studying enzymes in nonaqueous media? *Phil. Trans. R. Soc. Lond. B Biol. Sci. 359*, 1287– 1297.

[16] Lousa, D., Baptista, A. M., Soares, C. M. (2013). A molecular perspective on nonaqueous biocatalysis: Contributions from simulation studies. *Phys. Chem. Chem. Phys. 15*, 13723-13736.

[17] Carrea, G., Riva, S. (2000). Properties and synthetic applications of enzymes in organic solvents, *Angew. Chem. Int. Ed. 39*, 2226-2254.

[18] Micaelo, N. M., Soares, C. M. (2007). Modeling hydration mechanisms of enzymes in nonpolar and polar organic solvents, *FEBS J. 274*, 2424–2436.

[19] Clark, D. S. (2004). Characteristics of nearly dry enzymes in organic solvents: implications for biocatalysis in the absence of water, *Phil. Trans. R. Soc. Lond. B Biol. Sci. 359*, 1299–1307.

[20] Serdakowski, A. L., Dordick, J. S. (2007). Enzyme activation for organic solvents made easy, *Trends Biotechnol. 26*, 54-48.

[21] Rariy, R., Klibanov, A. M. (1997). Correct protein folding in glycerol, *Proc. Natl. Acad. Sci. U.S.A. 94*, 13520-13523.

[22] Bull, H. B. (1944). Adsorption of water vapor by proteins. *J. Amer. Chem. Soc. 66*, 1499-1507.

[23] Luscher-Mattli, M., Ruegg, M. (1982). Thermodynamic functions of biopolymer hydration. I. Their determination by vapor pressure studies, discussed in an analysis of the primary hydration process. *Biopolymers 21*, 403-418.

[24] Luscher-Mattli, M., Ruegg, M. (1982). Thermodynamic functions of biopolymer hydration. II. Enthalpy-entropy compensation in hydrophilic hydration process. *Biopolymers 21*, 419-429.

[25] Bone, S. (1987). Time-domain reflectometry studies of water binding and structural flexibility in chymotrypsin. *Biochem. Biophys. Acta. 916*, 128-134.

[26] Hnojewyj, W. S., Reyerson, L. H. (1961). Further studies on the sorption of H_2O and D_2O vapors by lysozyme and the deuterium-hydrogen exchange effect. *J. Phys. Chem. 65*, 1694-1698.

[27] Sirotkin V. A., Korolev D. V. (2005). Effect of acetonitrile on the hydration of human serum albumin films: a calorimetric and spectroscopic study. *Thermochim. Acta, 432*, 246-253.

[28] Sirotkin, V. A., Khadiullina, A. V. (2011). Hydration of proteins: excess partial enthalpies of water and proteins. *J. Phys. Chem. B*, *115*, 15110-15118.

[29] Sirotkin, V. A., Komissarov, I. A., Khadiullina, A. V. (2012). Hydration of proteins: excess partial volumes of water and proteins, *J. Phys. Chem. B*, *116*, 4098-4105.

[30] Borisover, M. D., Sirotkin, V. A., Solomonov, B. N. (1995). Isotherm of water sorption by human serum albumin in dioxane: Comparison with calorimetric data. *J. Phys. Org. Chem. 8,* 84-88.

[31] Sirotkin, V. A., Borisover, M. D., Solomonov, B. N. (1995). Heat effects and water sorption by human serum albumin on its suspension in water-dimethyl sulfoxide mixtures. *Thermochim. Acta, 256*, 175-183.

[32] Sirotkin, V. A. (2005). Effect of dioxane on the structure and hydration-dehydration of α-chymotrypsin as measured by FTIR spectroscopy. *Biochim. Biophys. Acta*, *1750*, 17-29.

[33] Kocherbitov, V., Arnebrant, T. (2010). Hydration of Lysozyme: The Protein-Protein Interface and the Enthalpy-Entropy Compensation. *Langmuir, 26*, 3918-3922.

[34] Foss, J. G., Reyerson, L. H. (1958). The sorption of water vapor by lyophilized ribonuclease. *J. Phys. Chem. 62,* 1214-1216.

[35] Amberg C. H. (1957). Heats of adsorption of water vapor on bovine serum albumin. *J. Amer. Chem. Soc, 79*, 3980-3984.

[36] Almog R., Schrier E. E. (1978). The dependence of the heat of solution of lyophilized solid ribonuclease A on water content. *J. Phys. Chem, 82*, 1701-1702.

[37] Smith A. L., Shirazi H. M., Mulligan S. R. (2002). Water sorption isotherms and enthalpies by lysozyme using the quartz crystal microbalance/heat conduction calorimeter. *Biochim. Biophys. Acta, 1594*, 150-159.

[38] Kocherbitov V., Arnebrant T., Söderman O. (2004). Lysozyme-water interactions studied by sorption calorimetry. *J. Phys. Chem. B, 108*, 19036-19042.

[39] Sirotkin, V. A., Zinatullin, A. N. Solomonov,B. N., Faizullin, D. A., Fedotov, V. D., (2002). Interaction enthalpies of solid bovine pancreatic chymotrypsin with organic solvents: comparison with FTIR-spectroscopic data. *Thermochim. Acta, 382*, 151-160.

[40] Fersht, A. (1999). *Structure and Mechanism in Protein Science: A Guide to Enzyme Catalysis and Protein Folding;* Freeman & Co: New York.

[41] Lehninger, A. L.; Nelson, D. L.; Cox, M. M. (1993). *Principles of Biochemistry*; Worth: New York.

[42] Perrin, D. D., Armarego, W. L. F., Perrin, D. R. (1980). *Purification of Laboratory Chemicals*, Oxford: Pergamon Press.

[43] Kogan, V. B., Fridman, B. N., Kafarov, V. V. (1966). *Liquid–Vapor Equilibrium*. Moscow: Nauka.

[44] Bell, G., Janssen, A. E. M., Halling, P. (1996). Water activity fails to predict critical hydration level for enzyme activity in polar organic solvents: Interconversion of water concentrations and activities. *Enzym. Microb. Technol. 20*, 471-476.

[45] Nikolskii, B. P. (1963). *Chemist's Handbook,* Leningrad, Russia: Goskhimizdat.

[46] Greenspan, L. (1977). Humidity fixed points of binary saturated aqueous solutions. *J. Res. Natl. Bur. Stand.*, Sect. A *81*, 89.

[47] Sirotkin, V. A., Kuchierskaya, A. A. (2017). α-Chymotrypsin in water-ethanol mixtures: Effect of preferential interactions. *Chem. Phys. Lett. 689*, 156-161.

[48] Sirotkin, V. A., Kuchierskaya, A. A. (2017). α-Chymotrypsin in water-acetone and water-dimethyl sulfoxide mixtures: Effect of preferential solvation and hydration. *Proteins: Functions, Structure and Bioinformatics, 85*, 1808-1819.

[49] Khurgin, Y. I., Sherman, F. B., Tusupkaliev, U. (1978). Temperature dependence of chymotrypsinigen A hydration in dynamic conditions. *Mol. Biol. 12*, 572-579.

[50] Sirotkin, V. A., Kuchierskaya, A. A. (2017). Preferential solvation/hydration of α-chymotrypsin in water-acetonitrile mixtures. *J. Phys. Chem. B. 121*, 4422-4430.

[51] Sirotkin, V. A., Kuchierskaya, A. A. (2017). Lysozyme in water-acetonitrile mixtures: Preferential solvation at the inner edge of excess hydration. *J. Chem. Phys. 146*, 215101-8.

In: Acetone
Editors: Irenka Kozłowska et al.
ISBN: 978-1-53613-298-4

Chapter 4

ACETONE IN THE CHARACTERIZATION AND UTILIZATION OF LIGNOCELLULOSIC BIOMASS: LIGNIN

Biljana M. Bujanovic*, PhD
Department of Paper and Bioprocess Engineering,
College of Environmental Science and Forestry,
State University of New York at Syracuse, NY, US

ABSTRACT

In recent times, lignocellulosic biomass (LCB) was mainly used to produce cellulosic pulp for paper and other cellulose products, such as cellulose derivatives. The pulping industry relies on inorganic agents, such as sodium hydroxide and sodium sulfide, that are combined to make the "white liquor" utilized in the kraft process, the dominant chemical pulping process used today.

Organic solvents are rarely used for pulping, even though organosolv processes have numerous advantages over kraft pulping, including better selectivity resulting in higher yields and less negative environmental

* Corresponding Author Email: bbujanovic@esf.edu.

impact. One important example of an organosolv process, the Alcell process, is based on ethanol, but acetone has been suggested for use as well. The Alcell process was implemented in a pilot plant facility based on hardwoods for about a decade at the end of XXth century producing attractive byproducts including lignin. However, organosolv pulping has not been broadly used for various reasons, including the inferior strength properties of organosolv pulp when compared to those of kraft pulp and negative results of economic analysis based on comparisons with larger scale kraft pulping operations. With increasing concerns over the depletion of fossil-based resources and the adverse effects that the combustion of fossil fuels may have on the global climate, there is a renewed interest in lignocellulosic biomass (LCB) and studies are underway to utilize all of the LC constituents, in addition to cellulose, to produce energy, chemicals, and materials in second generation biorefineries. Whereas starch- and sugar-based biomass such as corn and sugar cane are feedstocks for first generation biorefineries; lignocellulosic biomass such as wood agriculture residues and grasses provide the feedstocks for second generation biorefineries.

In this context, the processes that were used in the past should be re-examined and improved and new ones should be developed (for example the process of acetone-butanol-ethanol production by fermentation of sugars recovered by hydrolysis of lignocellulosic polysaccharides). The inorganic agents that are commonly used in conventional pulping processes may also be used in second generation biorefineries to decrease the recalcitrance of LCB in alkaline or acid based pretreatment processes, but organic solvents, especially renewable solvents such as ethanol and acetone, are expected to play a more prominent role in second generation biorefineries than in traditional pulping processes.

The part that acetone may play in increasing the use of lignocellulosic biomass through exploitation of all constituents, especially lignin, is examined here. Examples will include the use of acetone in studies of the chemical composition of LCB and the utility of acetone in organosolv pulping and pretreatment of LCB. Acetone is routinely used in solvent extraction procedures because its solubility parameter is compatible with that of lignin, making it an excellent choice for use in studies of the fractionation and purification of lignin as well as the production of different sizes of lignin, including nanoparticles, for higher value applications of lignin.

With the commercialization of biorefineries it is expected that acetone will continue playing an important role in the processing of lignin, the most abundant natural aromatic polymer on earth with immense potential to reduce dependence on fossil-based aromatic products.

1. INTRODUCTION

There is a renewed interest in lignocellulosic biomass with increasing concerns over the depletion of fossil-based resources and the negative effects which the combustion of fossil fuels may have on the global climate. Studies to utilize constituents of lignocellulosic biomass to produce a broad range of products, including energy, in second generation biorefineries are well-advanced. Unlike the starch-and sugar-based biomass, such as corn and sugar cane that are feedstocks for first generation biorefineries, second generation biorefineries rely on lignocellulosic biomass such as wood, agriculture residues, and grasses that cannot be utilized by food and feed industry [1]. The inorganic agents commonly used in conventional pulping processes will continue to be used in second generation biorefineries, mainly to decrease the recalcitrance of lignocellulosic biomass in processes such as alkaline or acid based pretreatments [2], however, organic solvents, especially renewable solvents including acetone, are expected to play a prominent role in the various operations of lignocellulosic biorefineries, including pretreatment processes [3]. Acetone is also frequently used in selective extraction experiments in studies of lignin fractionation [4] and in solvent-shifting experiments in the production of nanosize lignin particles [5]. Acetone and other organic solvents may be useful in different operations to process the separated LCB constituents into higher-value products, including the production of aerogels [6]. Furthermore, organic solvents have been used for the extraction of lignocellulosic biomass in fundamental studies such as the determination of chemical composition. As a relatively benign solvent in regard to its toxicity and environmental hazard [7] (acetone is on the list of Pfizer preferred solvents) [8], acetone has been increasingly used for this purpose [9]. Known for its ability to extract phenolic compounds, acetone has been frequently suggested as a solvent of choice for extraction of bioactive phenolic compounds from knotwood [10, 11]. The goal of this review is to report both the current and potential uses of acetone as an attractive solvent in advancing exploitation of lignocellulosic biomass as an alternative for fossil-based raw materials.

2. Acetone in Analysis of Chemical Composition of Lignocellulosic Biomass

Lignocellulosic biomass has the ability to store solar energy in the chemical form of its constituents. LCB includes trees (hardwoods, softwoods), energy crops like short-rotation wood species (e.g., willow, paulownia/hardwoods) and grasses (e.g., miscanthus, switchgrass), and agriculture residues (e.g., wheat straw, corn stover). Polysaccharides (cellulose and hemicelluloses) and lignin are the most abundant structural LC constituents. The structures of cellulose; acetylated 4-O-methyl glucuronoxylans, the most abundant hemicellulose in hardwoods; acetylated (galacto)glucomannans, the most abundant hemicellulose in softwoods; and lignin, as proposed to be typical of hardwoods [12] are shown in Figures 1-4, respectively. The presented structures illustrate a decrease in the order of uniformity in organization of structural polymer constituents.

Cellulose is the most uniformly organized polymer: a highly crystalline homopolysaccharide consisting of β-D-glucopyranose units connected to each other by β(1→4) glycosidic bonds in repeating blocks of cellobiose (Figure 1). The hemicelluloses are less organized branched hetero-polysaccharides with relatively regular repeating sequences, acetylated (arabino)glucuronoxylans are abundant in angiosperms (hardwood xylans, Figure 2) and acetylated (galacto)glucomannans) are abundant in softwoods (Figure 3). Lignin is a randomly organized three-dimensional aromatic polymer (hardwood lignin, Figure 4). Low-molecular weight organic extractives such as aliphatic, phenolic and terpenoid compounds are the most versatile in nature although these minor or "accessory compounds" are only present in low amounts [13]. LCB represents the major source of renewable carbon on earth with an estimated total global annual production of more than 10 billion tonnes [14]. The LC constituents can be transformed to materials, chemicals, fuels, and energy to replace non-renewable fossil resources [1] so there is a strong interest in developing second generation biorefineries based on LCB. Doing so could help to secure the starch- and sugar-based biomass feedstock of the first generation biorefineries for the food and feed industry. Table 1 shows chemical composition data of selected types of LCB [15-22].

The main polysaccharide (mainly cellulose but also hemicelluloses) constituents have been targeted for the production of fermentable sugars in designing biorefineries. [1] In addition to the reported constituents, grasses and agriculture residues contain proteins in relatively low amounts, for example, de Vrije et al., (2002) measured the protein content in miscanthus at 1.3% based on oven-dry biomass (OD basis) [22] whereas Huijgen et al., (2010) reported the nitrogen content in wheat straw of 0.2% [23] representing 1.25% proteins (based on the nitrogen to protein conversion factor of 6.25).

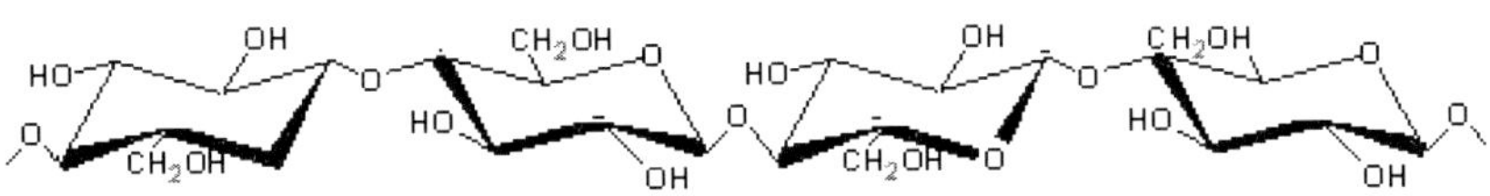

Figure 1. Structure of cellulose.

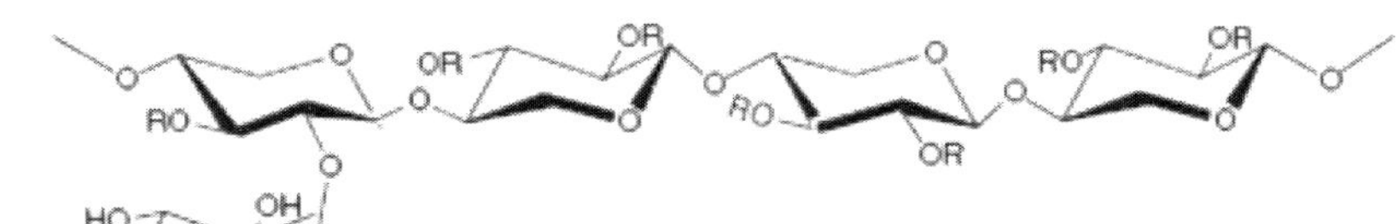

Figure 2. Structure of O-acetyl-4-O-methylglucuronoxylan in hardwood species.

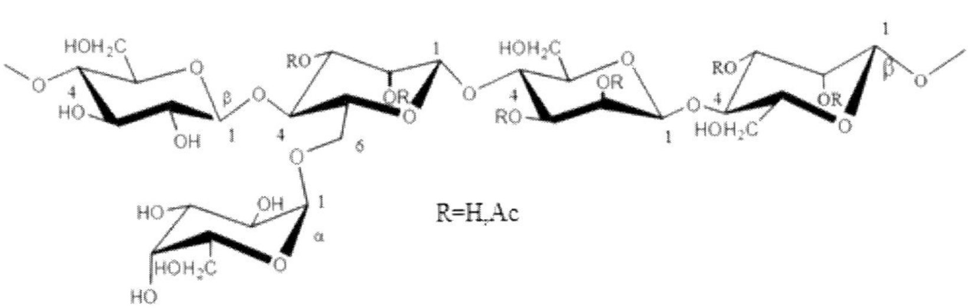

Figure 3. Structure of O-acetyl-(galacto)glucomannan in softwood species.

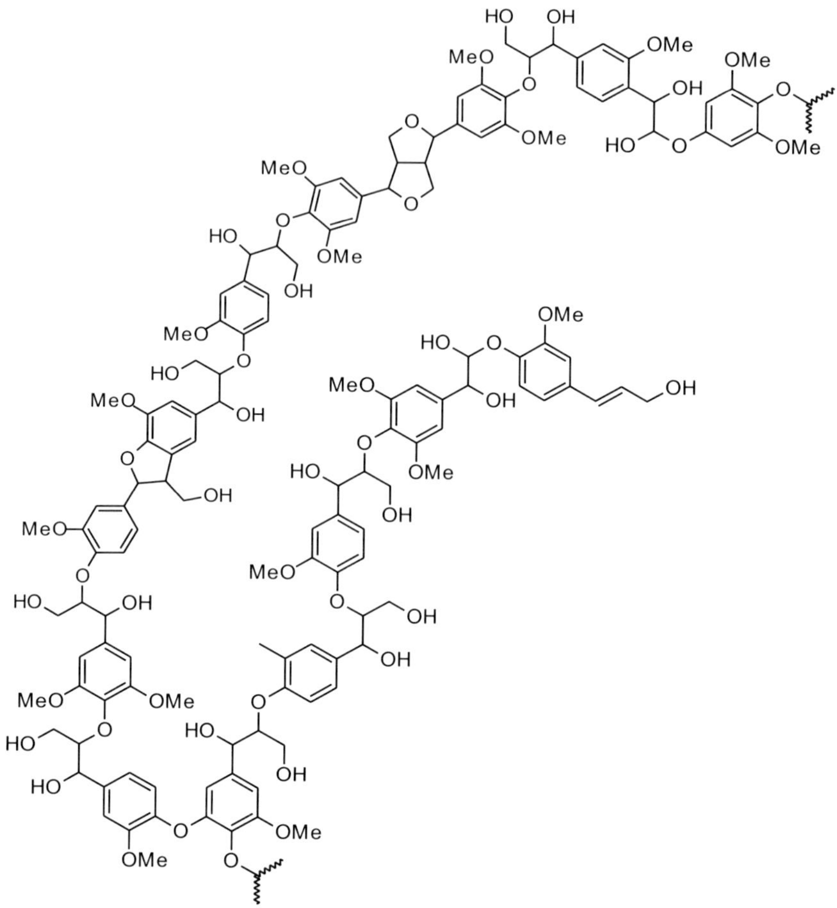

Figure 4. Lignin polymer model for a hardwood lignin, redrawn from Ralph et al., 2007 [12].

During analysis of the chemical composition of LCB, various organic solvents are used to determine the content of extractives [9] as shown in Table 1. An ethanol-benzene/toluene (1:2) mixture has been traditionally used in extraction studies. Benzene belongs to the category of carcinogenic/mutagenic/reprotoxic compounds (CMR, category 1 carcinogenic), is toxic to humans and the environment and classified as hazardous airborne pollutant in the US (HAP) [8]. Acetone is included on the list of Pfizer's preferred solvents [8], has fewer health hazards, is more environmentally favorable (Environmental Health Safety, EHS assessment per Capello et al., 2007) [7] and is increasingly used as an alternative solvent [24, 25] with a broad range of extracted compounds, including terpenoids (monoterpenoids, resin acids, sterols), aliphatic compounds (fats, fatty acids), and phenolic compounds (lignans) [9].

Methods of determination of extractives that are soluble in acetone are described in the TAPPI and SCAN standards [26, 27]. Experiments following these standards are most commonly conducted in Soxhlet and Soxtec type extractors and last typically 4 to 8 hours. Accelerated solvent extraction at high temperature and pressure has been developed to save time and reduce the amount of solvent required [9].

Table 1. Chemical composition of selected lignocellulosics

Lignocellulosic Biomass	**Constituent in % dry biomass**							**Ref**
	Cellulose	Hemicelluloses**			Lignin	Extractives***	Ash	
		X (Xylose)	GM	Total				
Scots pine *Pinus sylvestris* Gymnosperm	40.0	8.9	16	-	27.7	3.5 (DCM, E)	-	15
Sugar maple *Acer saccharum* Angiosperm	40.7	23.6	3.7	-	25.2	2.5 (DCM, E) 3.2 (AW) 2.8 (ET) 2.4 (E) DCM (0.6)	0.3	15-17
Willow *Salix viminalis** *Salix alba* Angiosperm	44.1 49.6	- 26.7	- -	21.8 -	20.4 22.7	3.6 (EB) -	2.1 0.3	17 18
Wheat straw *Triticum sp.* Angiosperm	32.0	26.1	-	26.9	17.7 17.7 19.1	2.7 (A) 4.2 (EB)	6.6 6.8	19-21
Miscanthus *Miscanthus sp.* Angiosperm	38.2	(19)	-	24.3	25.0	4.2 (ET)	2.0	22

* With bark; ** X-xylans; GM-glucomannans; *** extractives determined in the following organic solvents: DCM-dischloromethane; E-ethanol; AW-acetone-water (9:1); ET-ethanol-toluene (1:2); EB-ethanol-benzene (1:2); A-acetone

An acetone-water (9:1) mixture is also used as a powerful extraction solvent [16, 19] which in addition to typical acetone-soluble compounds can remove more of the hydrophilic compounds, such as sugars and phenyl glycosides. To determine the lipophilic extractives in this case, the acetone-water extract can be liquid-liquid extracted with a nonpolar solvent, such as hexane [9].

3. Acetone in Extraction of Knotwood

Knotwood is a tissue located in the branch bases inside the tree stems with a noticeably higher content of extractives than stemwood in both hardwoods and softwoods. This tissue is particularly rich in hydrophilic extractives soluble in acetone-water (95:5). In a comprehensive study of seven hardwood and fourteen softwood species, Pietarinen et al., (2006) found that the content of hydrophilic extractives in knotwood ranges from 2% to 29%, averaging 5.3% in hardwood and 8.4% in softwood, whereas the average extractive content of the stemwood of the eleven species of hardwoods and nine species of softwoods tested is typically 2.7% and 2.8%, respectively, when measuring total extractives by dichloromethane followed by ethanol. [13] Acetone-water extractives of hardwood and softwood knotwood are abundant in polyphenols, specifically lignans and stilbenes in softwoods and flavonoids in hardwoods (Figure 5).

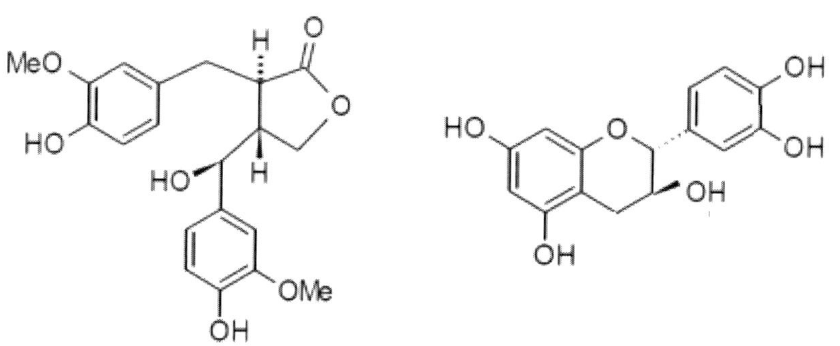

Figure 5. Lignan: Hydroxymatairesinol (lignan) and Flavonoid: catechin (flavan-3-ol).

The knotwood usually contains 50 to 100 times more polyphenols than the stemwood [10]. Polyphenols such as lignans and flavonoids exhibit excellent radical scavenger activity, i.e., they are strong antioxidants [10,

11, 28, 29]. They have also been suggested to possess antimicrobial, antifungal, and antiviral properties, and hence can be used in cosmetic, biocide and food applications since the dietary intake of polyphenols is considered beneficial for human health. Moreover, studies have shown that natural polyphenols produce some favorable effects in treatment of various diseases, including cancer, heart diseases, and stroke [28, 29] .

In this context, separation of knotwood prior to wood processing in the pulp and biorefinery industries, and the subsequent extraction of knotwood with acetone-water (95:5) would provide polyphenolics for various applications. This approach is especially attractive because processing of knotwood in industrial operations has always been problematic due to characteristic high density, high extractive content, and short fibers [30]. Therefore, knotwood separation would result in increasing both the efficiency of industrial operations and quality of products.

4. Acetone in Lignin Fundamental Studies

The emergence of lignin as a hydrophobic phenylpropanoid polymer is associated with the development of vascular plants (*Tracheophytes*) about 430 million years ago. [31] It is attributed to a vital need for biomechanical support of stems, cell reinforcements, and improvement of both water conduction and resistance to degradation by pathogens. Lignin is derived primarily from *p*-hydroxycinnamyl alcohols and results from the radical-coupling polymerization reactions in the cell wall, after polysaccharides have been laid down. During polymerization, coumaryl, coniferyl and sinapyl alcohols are converted to phenylpropanoid lignin units; *p*-hydroxyphenylpropanoid units; H, guaiacyl units; G, and syringyl units; S, respectively (Figure 6).

Different contribution of *p*-hydroxycinnamyl alcohols in the biosynthesis of softwood (G-lignin), hardwood (SG-lignin), and grass (HSG-lignin) lignins causes differences in the lignin structure, including the type and content of different bonds between phenylpropanoid units, and functional groups (Figures 4, 7). In addition, lignins may contain characteristic end groups, such as cinnamaldehyde units and γ-acetates (kenaf bast fiber lignin), γ-*p*-coumarates (grass lignins), and γ-*p*-hydroxybenzoates (poplars, willows) [12, 32].

Lignin provides a matrix in which polysaccharides are embedded and cross-linked, resulting in a naturally occurring composite material that imparts strength and rigidity to lignocellulosic plants. In addition, the lignocellulosic biomass is reinforced by close association of lignin with polysaccharides (lignin-carbohydrate complex; typical bonds illustrated in Figure 8) [17].

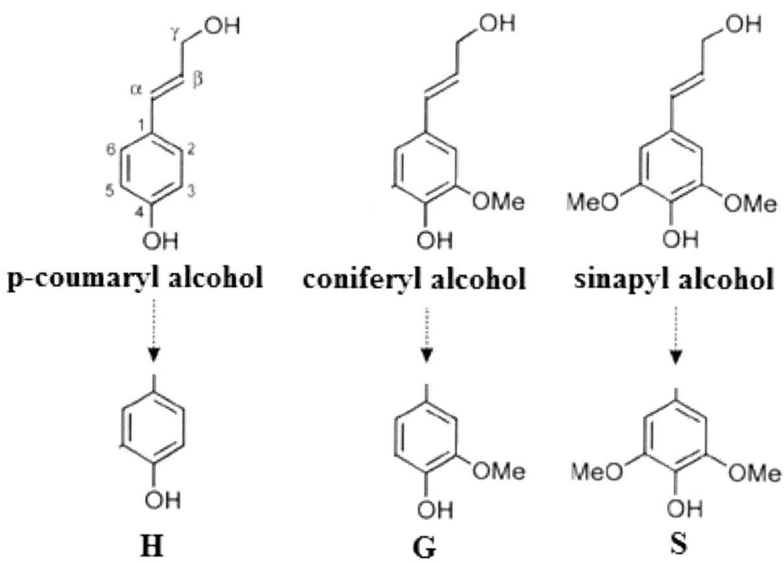

Figure 6. Lignin precursors and units.

A specific type of lignin-carbohydrate complex may be found in grasses where ferulates incorporated in lignin make bridges between lignin and xylans (Figure 9) [12, 33, 34]. These factors, along with the high degree of polymerization and crystallinity of cellulose as a major constituent of lignocellulosics (Table 1), are critical in building an intrinsic resistance toward biological deconstruction which is commonly referred to as recalcitrance of LCB.

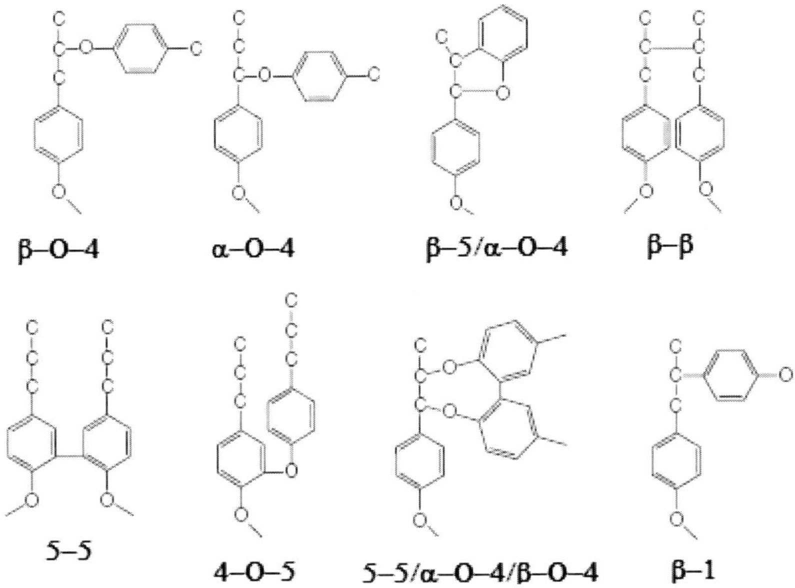

Figure 7. Common linkages between phenylpropanoid units in lignin [12].

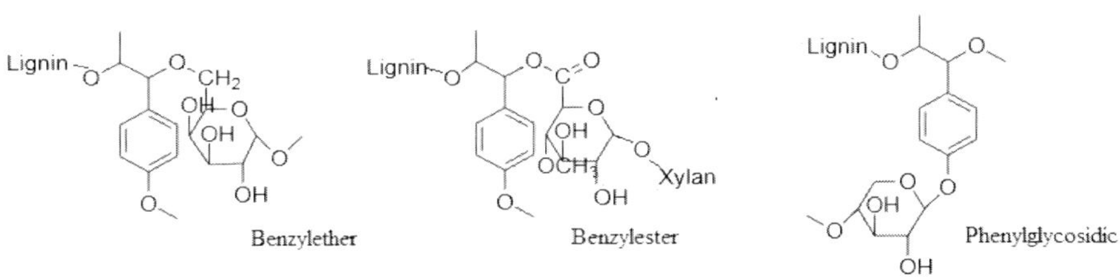

Figure 8. Some proposed chemical bonds between lignin and carbohydrates [17].

One of the main challenges in developing second generation biorefineries is reducing this recalcitrance to improve both fractionation into individual constituents and the accessibility of cellulose to hydrolytic enzymes. Softwoods have been confirmed as more recalcitrant than angiosperms most likely due to the higher content of lignin (Table 1) and the more resistant nature of softwood G-lignin in comparison to angiosperm (H)SG-lignin [2, 3, 14, 35, 36].

Since lignin is the most abundant natural aromatic polymer, accounting for ~30% of all of the organic carbon on earth [12], valorization of lignin in the pulp, paper, and biorefinery industries would contribute to their sustainability and profitability [36-38].

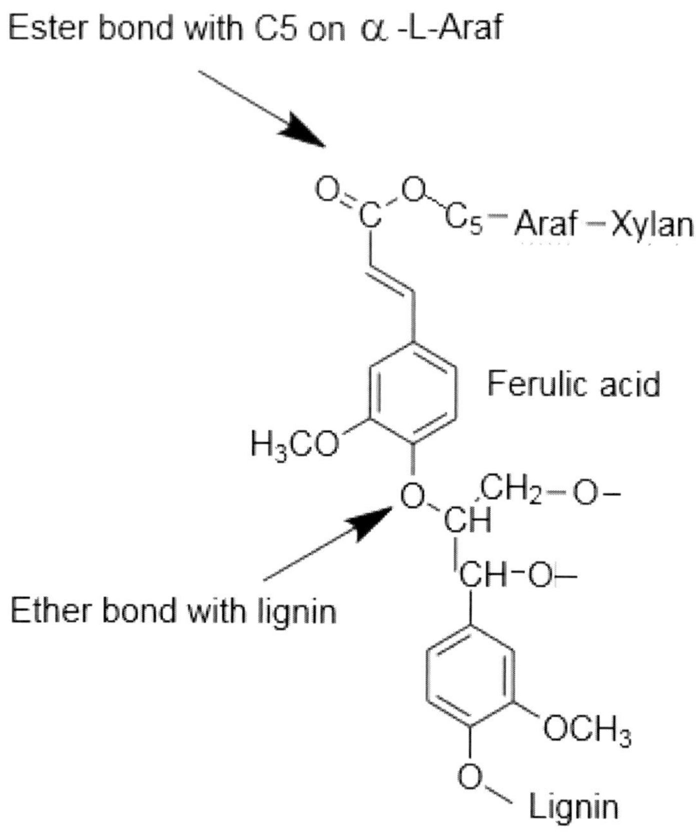

Figure 9. Proposed role of ferulic acid as a bridge between xylans and lignin in grasses [33, 34].

Currently, the annual world production of lignin by kraft pulping, the predominant chemical pulping process, is estimated to be 70 million tonnes, while just 0.1 million tonnes of this lignin is isolated and commercially available [39]. Only a minor amount of lignin is being recovered because the kraft process depends on the efficient regeneration of chemicals that incorporates the combustion of the organic material remaining in the residual "black liquor" after pulping, predominantly lignin. Analysis shows that the approximate value of 1 kg of lignin used as fuel is $0.04-0.18 vs. lignin used in chemical conversion at $1.08-1.10 [36, 37]. With the development of second generation biorefineries in the US, the amount of lignin that will be generated as a byproduct is estimated at 62 million tons by 2022 [5]. The amount of lignin produced per year in an ethanol commercial facility is expected to range from 100 to 200 thousand tons [38]. Advancing the utilization of lignin for higher-value applications requires a better understanding of its nature, therefore finding accurate methods of quantification, isolation and recovery, and characterization is of critical importance.

4.1. Quantitative and Qualitative Lignin Studies

Quantitative lignin analyses are commonly conducted following gravimetric and spectral methods. The most broadly used gravimetric method is the Klason procedure which is based on acid hydrolysis of polysaccharides with most of the lignin remaining insoluble. It is combined with UV-spectral analysis of the solution at aromatic maximum of 200-205 nm to take into account a small portion of acid-soluble lignin. To avoid drawbacks of the time-consuming two-step Klason procedure, the acetyl-bromide method based on UV-spectral analysis has been recommended; UV-absorption of the solution resulting from the biomass dissolution in acetyl bromide-acetic acid mixture is measured at characteristic lignin absorption maximum at 280 nm [40]. Qualitative lignin analyses may be performed on lignin isolated from finely ball-milled biomass either by extraction with dioxane-water (milled wood lignin), by enzymatic hydrolysis followed by extraction with dioxane-water (cellulolytic enzyme lignin) or by a combination of mild enzymatic hydrolysis and mild acidolysis (enzymatic mild acidolysis lignin) [19, 40-50]. Prior to quantification or isolation of lignin for qualitative studies, biomass must be extracted to remove interfering phenolic extractives. This mandatory pre-extraction step is commonly conducted with ethanol-benzene (1:2) [18, 46, 47, 51] and ethanol-toluene (1:2), [44, 45] although Björkman, who designed the first procedure for isolation of milled wood lignin, recommended a two-step extraction with ethanol-benzene (1:2) and 95% ethanol [41, 49]. In accordance with acetone's ability to efficiently remove lignin-like phenolic compounds of low-molecular weight and its favorable environmental and health properties [7-9], acetone and acetone-water solvents are increasingly used to extract biomass prior to lignin quantitative and qualitative studies, i.e., ball-milling for lignin isolation. Numerous research groups have used acetone or acetone-water for the biomass pre-extraction in their lignin studies since 1981 [16, 19, 24, 25, 42, 43, 48, 50, 52, 53].

Following isolation from finely ground biomass, lignin is characterized by different methods, some of which require lignin in a solution for analysis. Acetone has been recognized as a good solvent for isolated lignins, equivalent in solvent power to dioxane, although it is unable to completely dissolve alkali lignins, kraft and soda lignins. This is in accordance with the Schuerch's definition of good lignin solvents stating that lignin solvents should have the Hildebrand solubility parameter, δ_H (δ_H in $(calcm^{-3})^{1/2}$) of ~11 (22.5 $MPa^{1/2}$) and a hydrogen bonding capacity, $\Delta\mu$, of 0.14 μm or higher [54]. The Hildebrand solubility parameter, δ_H, is an indicator of the intermolecular interactions in solutions of polymers which depends on the vaporization enthalpy and the molar volume of the polymer ($\delta_H = (E/V)^{1/2}$). Polymers exhibit the best solubility in solvents with δ_H closest to their own, especially when these solvents are characterized by higher hydrogen-bonding capacity. Hydrogen-bonding capacity is expressed as a shift in oxygen-deuterium band in the IR spectrum when hydrogen-bonding solvents are mixed with CH_3OD [54]. Acetone and dioxane are characterized by the same δ_H of 10 (20.5 $MPa^{1/2}$) and $\Delta\mu$ - value of 0.14 μm, whereas pyridine is a better lignin solvent with δ_H of 10.7 (21.9 $MPa^{1/2}$) and $\Delta\mu$-value of 0.27 μm. However, dioxane and pyridine are both on the list of category 3 CMR compounds, and dioxane is also classified as HAP in the US [8]. Therefore, the use of acetone for lignin dissolution in lignin studies may provide appreciable health benefits.

Acetone is used for the solubilization of lignin in liquid-state NMR spectral studies that have contributed tremendously to elucidating lignin structure throughout the years. Qualitative, semi-quantitative, and quantitative proton, carbon, and two-dimensional NMR studies have helped to clarify a broad range of lignin features, including the contribution of H-, G-, and S-units; the type and contribution of lignin interunit bonds/substructures; the presence and content of specific functional groups; and units that did not originate from *p*-hydroxycinnamyl alcohols. NMR spectroscopy has been particularly instrumental in the development of methods to make compositional changes in plant lignins by manipulating the biosynthetic pathway genes to affect the success in destruction of lignocellulosic species for more efficient utilization. [55] Assignments of chemical shifts in these studies are based on data collected in a considerable amount of work on lignin and cell wall model compounds [56]. Acetone-d_6 has been recommended as a solvent of choice in studies on underivatized lignins with the intention to correlate observed shifts with the shifts of lignin model compounds [55]. It should be noted that the addition of water (δ_H of 23.4 (47.9 $MPa^{1/2}$)) to acetone increases the solubility of lignin due to the increased hydrogen capacity of the mixture. Therefore, an acetone-d_6-D_2O mixture (~9:1) (δ_H of 12.0 (24.6 $MPa^{1/2}$) [4] is also used in NMR studies [32, 33, 50, 55, 57].

5. Acetone in Delignification

5.1. Pulping

The dominant chemical pulping process today is kraft pulping with the total world production of ~133 million tonnes (>75% of the total wood pulp production in 2015). The US production of kraft pulp, bleached and unbleached, exceeded 41.8 million tonnes or approximately 31.5% of the total world production in 2015 [58]. This process uses a solution of inorganic chemicals comprised of sodium hydroxide and sodium sulfide called white liquor. Kraft pulping produces a superior strength pulp and has other advantages, such as efficient chemical regeneration which includes combustion of the spent black liquor with concurrent energy production and a high tolerance for variation in the quality of the raw materials used and the presence of some bark. In spite these benefits, alternative delignification processes have been explored for various reasons, including the need to protect the environment by eliminating emissions of reduced forms of sulfur, lessening water usage and increasing pulp yields to reduce the consumption of biomass. The kraft process has a relatively low delignification selectivity, especially in the residual phase of delignification [13]. Organosolv pulping is attractive as a sulfur-free alternative to kraft pulping because it is based on use of biorenewable solvents, such as ethanol, methanol, or butanol [20, 59, 60] that can be produced in second generation biorefineries [1]. Delignification has been performed in autocatalyzed, and acid- and alkali-catalyzed processes [61]. The yields are comparable or better than kraft pulp yields at the same kappa number (an indirect measure of the lignin content in pulps) and the bleachability is improved [59, 62], but the papermaking properties of organosolv pulps, especially those produced in acid-catalyzed conditions are inferior to kraft pulps [61]. In this context, the sulfur-free organosolv processes may be beneficial for the production of dissolving pulp [59] and nanocellulose [63]. Organic solvents are considered advantageous for lignin removal because of their ability to dissolve lignin fragments and lessen lignin condensation reactions that are dependent on the pH of the system and solvent's chemical properties [64]. Better delignification results have been reported in organosolv experiments conducted on hardwoods and grasses (angiosperms) than on softwoods, a fact that can be attributed to greater resistance of G-lignins as compared to (H)SG lignins. Lignin recovered from these processes is characterized by higher purity, and lower molecular weight and polydispersity in comparison to kraft lignin [37]. These features promote the use of organosolv lignin for higher value applications, such as production of carbon fibers [65].

Among the seven most investigated organosolv processes only Alcell and Organocell have been scaled up for commercialization [61]. The Alcell process is based on the use of aqueous ethanol primarily on hardwoods (ethanol-water, usually in a 5-to-5 volume ratio at 195°C) and functions in line with biorefinery concepts of careful fractionation of biomass for complete utilization of raw materials so that in addition to pulp, products include lignin, furfural, hemicellulose-based carbohydrates, and acetic acid. Ethanol is also recovered by distillation. The Alcell pulp can be used in a broad range of paper grades including tissue and towel. Alcell produces sulfur-free lignin of high purity that can be incorporated into a phenol-formaldehyde resin for use in a number of different industries including wood products (particleboards, plywood), rubber (tackifier), automotive (friction products such as brake pads), and paper (linerboard reinforcement). In addition, based on the Alcell lignin's natural antioxidizing potential, it can be used by the animal feed, rubber, and lubricant industries. Nevertheless, the Alcell process was only used commercially for only a short time from 1989 to 1996 in a facility New Brunswick, Canada that was closed for financial reasons [1, 61]. The Organocell process was designed for softwoods to produce fluff pulp (a type of pulp used in baby diapers) and was used in commercial pulp production in Germany for about a year in the early 1990s. The pulping is conducted in alkaline conditions in liquor containing methanol, sodium hydroxide, anthraquinone, and water. The process includes the stripping of methanol followed by the recapture of sodium hydroxide in a recovery boiler, similar to kraft pulping [61]. Organic material that was dissolved during the process, consisting mainly of lignin, is burned in the recovery boiler. In this regard the Organocell is not a typical organosolv process intended to promote complete utilization of biomass.

Organosolv delignification has been conducted using acetone in different processes, including acetone-water [66-68] and acetone-ethanol. [69] Moreover, delignification in a formic acid-acetone mixture [70] and oxygen delignification in acetone-water (AWO) [71, 72] have been shown efficient on both softwoods and hardwoods. It has also been demonstrated that organosolv processes benefit from different mild pretreatment processes, similar to those used in mechanical and chemical pulping processes, including hydrothermal treatment (autohydrolysis) and biodegradation [61]. The effects of different pretreatments on organosolv pulping processes based on acetone have been studied. For example, delignification in formic acid-acetone mixture was more efficient when performed on pine with introduced white-rot than on native pine. These results demonstrated that fungal pretreatment provides an increase in tensile strength of the produced formic acid-acetone organosolv pulps. This is a noteworthy result, as the main drawback of organosolv pulps is attributed to their low strength [73]. Pulping experiments have been conducted following different delignification methods on hydrothermally pretreated hardwoods (woodchips, 2 hours at 160°C). Even though an increase in delignification efficiency was observed for pretreated biomass in most of the investigated processes, including kraft, superior results were obtained for AWO delignification [16, 74-76]. The results obtained in relatively modest AWO conditions (30 mesh particles, 2 hours at 150°C, O_2 pressure 1.38 MPa, acetone-water (6:4) on a number of pretreated hardwood species indicated that the delignification degree may exceed 95%, a significant improvement compared to that for native species, which was as low as 50%. In contrast to these results, in organosolv experiments performed following an autocatalyzed ethanol-water (6:4) procedure (30 mesh particles, 2 hours at 195°C) the delignification degree for native and pretreated sugar maple was found to be approximately equal at ~80%, with no observed improvement due to prior hydrothermal treatment. It appears that a significant increase in free phenolic hydroxyl groups that is caused by the cleavage of lignin aryl-ether bonds during hydrothermal pretreatment facilitates oxygen delignification in acetone-water (AWO) but not delignification in ethanol-water [76].

There are many advantages of organosolv delignification processes in comparison to the dominant kraft pulping process, including the ease of recycling organic solvents, lower capital costs, and smaller scale production for economically efficient operation. Creating a revenue stream based on high quality lignin as a byproduct is very important in the context of complete utilization of lignocellulosic biomass to develop biorenewable products. However, organosolv pulping has not been broadly accepted due to various reasons such as inferior pulp strength compared to that of kraft pulp. The volatility of solvents also creates potential safety issues; including fire and explosion hazards [3, 7, 61].

5.2. Pretreatment

The successful implementation of secondary biorefineries vitally depends on deciphering the recalcitrance of lignocellulosic biomass and the development of efficient pretreatment technologies [2, 3, 14, 35]. Extensive research has been performed to evaluate pretreatments designed for one or more of these results: removing hemicelluloses and lignin; reducing cellulose crystallinity and degree of polymerization; and increasing porosity to facilitate the fractionation of lignocellulosic constituents and the processability of the cellulose-enriched residue, most importantly enzymatic hydrolysability. Physical, chemical, and biological pretreatment methods have been suggested [14]. Among these, autocatalyzed hydrothermal methods (autohydrolysis) using only water and heat, have been recommended frequently to create a cleaner, more environmentally benign process with a low corrosion risk for equipment and lower quantities of potential downstream inhibitors produced [77]. Hydrothermal methods have been shown efficient in pretreatments of angiosperms (hardwoods and grasses) in regard to the relatively selective removal of xylans with consequent increase in enzymatic hydrolysability and delignification [14, 16, 74, 75, 76]. In studies of chemical composition, an increase in extraction yields due to hydrothermal treatment has been detected in Soxhlet extraction experiments on native and hydrothermally treated hardwoods using different solvents [16]. In addition, for hydrothermally treated biomass the content of lignin decreases after solvent extraction, especially after acetone [25] and acetone-water (9:1) extraction [16] indicating the occurrence of delignification during extraction; a result that is obviously in conflict with the pre-extraction goal of removing low-molecular weight extractives, particularly the phenolics that can interfere with the subsequent determination of lignin. 2D NMR analysis confirmed that the acetone-water extracted material contains lignin and lignin-like compounds [50]. Therefore organic solvents must be carefully selected to avoid any delignification during pre-extraction for the purpose of lignin determination in hydrothermally pretreated lingo-cellulosics [16]. On the other hand, it has been proposed that acetone can be used to remove lignin and lignin-like compounds by extraction to promote further processing of hydrothermally pretreated biomass in the secondary biorefinery, e.g., enzymatic hydrolysis [50, 78, 79].

Organosolv processes inherently complement second generation biorefineries as they use solvents that are the designed products of biorefineries [1, 3, 14, 61]. The organic solvents and each lignocellulosic constituent can be recovered at the end of the process, a feature consistent with complete utilization of LCB as the ultimate goal of the second generation biorefinery. In organosolv pretreatment, the targeted degree of delignification is lower than in organosolv pulping, however the delignified product in this case is intended for further processing and is not required to meet the stringent strength requirements that are typical of paper industry. Hemicelluloses are also removed to a variable extent depending on the process conditions, leaving the cellulose-enriched product as an insoluble residue for further chemical or biological processing. The organosolv pretreatment does not usually require a catalyst because the organic acids released during the process serve as *in situ* catalysts, although both inorganic and organic acids can also be used. Organic solvents must be removed from the remaining cellulose-enriched product before biological processing, enzymatic hydrolysis and fermentation due to their inhibitory effect [3].

Acetone-based organosolv processes have been explored for the production of monosaccharides from lignocellulosics. A patent was issued in 1983 providing data on acid-catalyzed (phosphoric, sulfuric, hydrochloric acid), acetone-water delignification with simultaneous saccharification (hydrolysis of polysaccharides) of broad range of lignocellulosics, including Douglas fir and spruce, the most resistant softwoods [80]. Following the evaporation of acetone from the spent liquor, lignin can be separated from the carbohydrates by filtration. A significant advantage of the proposed process is a minimal loss of monosaccharides due to degradation. Hydroxymethyl furfural and furfural can also be produced by the dehydration of hexoses and pentoses, respectively, along with a range of other low-molecular weight organic compounds when the proposed treatment is conducted at higher temperatures for longer periods of time. Furfural is a degradation product of xylose from xylans, the most abundant hemicellulose in angiosperms (Table 1). A broad range of applications of furfural including production of polymers (Nylon 6) may provide an important revenue stream for biorefineries [36].

The use of an autocatalyzed acetone-water treatment has been explored as a pretreatment to improve access to cellulose and to facilitate the subsequent enzymatic hydrolysis of cellulose. Experiments performed on wheat straw demonstrated a high increase in cellulose hydrolysability due to this type of pretreatment; based on the total glucose available. The results revealed a glucose yield of 16% vs. 87% for untreated wheat straw vs. wheat straw pretreated in acetone-water (4:6) for 1 hour at 190°C [23]. The use of acid-catalyzed acetone-water (1:1) pretreatment (sulfuric acid, pH2) to produce glucose by enzymatic hydrolysis and ethanol by simultaneous saccharification and fermentation (SSF) has been studied on softwood *Pinus radiata*. These studies show that the enzymatic hydrolysis and fermentation yields depend on the combined effect of time and temperature during pretreatment and may produce glucose yields greater than 71%, based on the total glucose available in the pretreated pine, a sizeable increase for the most recalcitrant softwood species. The ethanol yields can reach 99.5% of theoretical yields in 72 hours of SSF at 40°C [81].

A two-step phosphoric acid-acetone process has been suggested to increase cellulose hydrolysability in different types of LCB including softwood Douglas-fir and angiosperms; hardwood poplar, corn stover, and switchgrass. In the first step, Zhang and coworkers [36] use concentrated phosphoric acid as an excellent cellulose solvent to achieve several goals including the dissolution of cellulose and hemicelluloses; a decrease in the degree of polymerization of cellulose; deacetylation of hemicelluloses; and the disruption of lignin-carbohydrate complex (60-40 mesh particle size, 50°C, 30-60 minutes). In the second step, acetone is added into the reaction vessel for various objectives, most importantly, the precipitation of the dissolved cellulose (and hemicelluloses) in the amorphous form that is intrinsically more hydrolysable than crystalline cellulose. Acetone also dissolves some of the lignin. The precipitate is washed with acetone and water before enzymatic hydrolysis, while the liquor is processed to recover lignin, acetic acid, acetone, and phosphoric acid. The glucose yields exceeded 95% after 24 hours of enzymatic hydrolysis for the angiosperms while lower yields of 75% were obtained in softwood, due to less effective removal of lignin. These are remarkable results, especially considering the relatively short enzymatic hydrolysis that was used in these studies (24 hours vs up to 72 hours in other studies). The proposed two-step process has been used on hemp hurds and rice straw, corroborating the previous results showing that this concept is highly effective at improving the yields of enzymatic hydrolysis [82, 83]. Glucose produced from rice straw was used in acetone-butanol-ethanol fermentation demonstrating that acetone, a biorenewable solvent suggested for use in operation of the second generation biorefinery, could also be produced in this fashion [83]. Historically, acetone-butanol-ethanol fermentation was one of the most important fermentation processes in starch- and sugar-based biorefineries at the beginning of the 20th century before acetone became a petrochemical product [84].

6. Acetone in Lignin Enhancement and Uses

Various challenges have been encountered in studies and attempts to use lignin for higher-value applications. They are closely related to the inherent variability of lignin between and within species and even within a single plant. As an example, lignin in wood has a variable structure depending on the morphological location within the plant cell. There are differences in lignin from the compound middle lamella vs. lignin in the secondary wall and variability is present in the lignin found in hardwood fibers vs. vessel cells [13]. In addition to the origin, the process used to isolate lignin from biomass critically influences its final structure and properties. Technical lignins are thus categorized by taking into account the factors related to the biomass and process origins and named accordingly (i.e., softwood kraft lignin, hardwood organosolv lignin). With implementation of different biorefinery processes, the technical lignins will include new types and different lignins that will result from different pretreatment and fractionation processes; for example, the residue after enzymatic hydrolysis will be a lignin-enriched material. Consequently the need for an industrially acceptable uniformity across different lignin types will become more urgent and methods of processing to help meet this goal will be required. The lignin features deserving special attention are purity (freedom from carbohydrates and inorganics), distinct functionalities, molecular weight and polydispersity High polydispersity of technical lignins is considered especially critical in efforts to make lignin-based functional polymers and materials [39]. In this context a range of lignin fractionation methods has been recommended by different research groups. These methods include: membrane separation; precipitation through the gradual decrease of pH in a lignin-alkali solution, where lignin that is soluble in alkaline conditions becomes increasingly insoluble with an increase in acidity; and sequential organic solvent extraction [85].

Lignin fractions of different molecular weights and functionalities have been produced in experiments which use various sequences of solvents with varying Hildebrand solubility parameters, δ_H $(calcm^{-3})^{1/2}$. The well documented lignin solvents which may be considered for this use include dimethyl sulfoxide, DMSO (δ_H 13.1 (26.6 $MPa^{1/2}$), dimethyl formamide, DMF (δ_H 12.2 (24.9 $MPa^{1/2}$), and tetrahydrofuran, THF (δ_H 9.5 (19.5 $MPa^{1/2}$) [85] (δ_H of DMSO, DMF, and THF per Vangeyete et al., (2004) [86]). Regardless, acetone is frequently used for lignin solubilization due to the high boiling temperature of DMSO and DMF (189°C and ~153°C, respectively in contrast to 56°C of acetone) and the fact that THF is more problematic for use than acetone based on EHS and life-cycle assessments [7].

Ropponen et al., (2011) examined softwood and hardwood kraft lignin, and birch and spruce organosolv lignin in sequential extraction experiments using solvents of increasing Hildebrand solubility parameter, diethyl ether (δ_H 7.5 (15.3 $MPa^{1/2}$)), diethyl ether-acetone (4:1), and acetone (δ_H 10.0 (20.5 $MPa^{1/2}$)) (δ_H per Schuerch (1952) [59]). Fractions of lignin isolated with acetone exhibited a more homogenous nature of narrow molecular distribution. The acetone soluble lignin fractions had a notably lower glass transition temperature (T_g) and higher heat capacity (ΔC_p) than that of the starting lignin or the lignin remaining as insoluble residues at the end of fractionation. These features of acetone soluble lignin fractions promote thermoplastic lignin properties stimulating interest in lignin for use in thermoplastic composites [87].

Comprehensive fractionation studies of five technical lignins were performed by Boeriu et al., (2014). The lignins were extracted using aqueous acetone of decreasing acetone-to-water ratio and therefore of increasing Hildebrand solubility parameter (9:1, δ_H 12.0 (24.6 $MPa^{1/2}$); 7:3 δ_H 15.8 (32.3 $MPa^{1/2}$); 5:5 δ_H 18.8 (38.5 $MPa^{1/2}$); 3:7 δ_H 21.1 (43.2 $MPa^{1/2}$); 1:9 δ_H 22.8 (46.6 $MPa^{1/2}$)) and by ethylacetate (δ_H 7.6 (15.6 $MPa^{1/2}$)). These studies showed that for all of the examined lignins, an ethylacetate and acetone-water (3:7) extraction produces a lignin that has lower molecular weight and dispersity and is also less condensed in comparison to starting (parent) lignin. Extraction yields showed a dependence on the type of lignin. Insoluble fractions that resulted from extraction with acetone-water of δ_H ≤18.8 (38.5 $MPa^{1/2}$) were of relatively higher molecular weights than the starting lignins. Lignin-carbohydrate complexes were isolated from all lignins by extraction with acetone-water (1:9). These studies included the calculation of Hildebrand solubility parameters, δ_H of examined lignins based on the molar ratios of H, G, and S units present in these lignins (determined from ^{31}P NMR data) and δ_H-parameter of these units: H, δ_H 14.1 (28.8 $MPa^{1/2}$), G δ_H 13.5 (27.7 $MPa^{1/2}$), S δ_H 14.2 (29.1 $MPa^{1/2}$). Accordingly the Hildebrand solubility parameter of five technical lignins was shown to range from δ_H 13.6 to 14.3 (27.8 – 29.3 $MPa^{1/2}$). The authors suggested that in addition to Hildebrand solubility parameters other factors should be considered when choosing an appropriate solvent such as the Hansen solubility parameters and the difference in the solubility parameters of the lignin and the solvent. [4] The Hansen solubility parameters are expressed as the cohesion energy densities for nonpolar (London) interactions, δ_D; permanent dipole interactions, δ_P; and hydrogen bonding interactions, δ_H. The Hansen solubility parameters that have been recommended for lignin are $\delta_D = 21.9$ $MPa^{1/2}$, $\delta_P = 14.1$ $MPa^{1/2}$, and $\delta_H = 16.9$ $MPa^{1/2}$. [88]

Studies aimed at isolation of narrow fractions from three different kraft lignins characterized by high polydispersity ranging from 4.5 to 10 were performed by Cui et al., (2014) who were able to develop a protocol for the fractionation of kraft lignin which includes the use of acetone [85]. Acetone extracts of three studied kraft lignins were used to produce lignin fractions by incremental addition of hexane as a nonpolar miscible solvent (δ_H 7.4 (15.1 $MPa^{1/2}$; identified as a nonlignin solvent by Schuerch (1952) [54]) which caused gradual lignin precipitation. The results demonstrated that by manipulating the amount of hexane in the acetone-hexane mixture the composition of produced lignin can be tailored and that nearly monodisperse lignin of low molecular weight with polydispersity ranging from 1.1 to 1.3 can be produced. The suggested protocol is a promising method to increase lignin uniformity as it yielded lignin fractions of almost the same properties, independent of the type of original kraft lignin. However, for industrial applications the use of high quantities of hexane which is classified as a HAP compound in the US is undesirable and alternative solvents such as less toxic heptane should be investigated [8].

Solvent shifting methods have been explored to produce lignin nanoparticles of superior properties (i.e., radical scavenging ability, UV-light blocking properties) in comparison to the original or parent lignin. Solvent shifting methods are used to create nanoparticles of organic compounds from their solution in a water-miscible solvent which is mixed with an excess amount of water as the antisolvent. The procedure is conducted by nanoprecipitation where the solution of an organic compound is added rapidly to water, or by flash precipitation in which water is added to supersaturate the organic compound dissolved in solution, or by introducing water into the system through dialysis [89-91]. Experiments have been conducted from lignin solutions in various lignin solvents including THF, ethylene glycol (δ_H 14.2 (29.1 $MPa^{1/2}$)) [54], which is extremely toxic, acetone, and acetone-water solvents. While the structural and thermal properties lignin have not been changed, the UV-absorption and antioxidation properties of nanoparticles are increased compared to parent lignin [5, 91-93]. An interesting method of the production of nanoscale lignin particles has been proposed by using acetone as a solvent and super-critical carbon dioxide (SC-CO_2) as an antisolvent. An acetone solution of poplar organosolv lignin was injected into the precipitation chamber filled with SC-CO_2 and acetone at high pressure, forming nanoscale lignin particles after the acetone had been carried away with SC-CO_2 and the pressure was decreased to atmospheric. With the decrease in lignin particle size from its original 10^3 to 10^5 nm to the mean particle size of 144 nm, the specific surface area increased more than eight times while the radical scavenging activity increased more than ten times [94]. Lignin nanoparticles have been studied for various potential applications to be implemented in agriculture, food, pharmaceutical, and the cosmetic industries. Possible uses include reinforcement of natural rubber composites and phenolic foams, stabilization of emulsions, adsorption of heavy metals in environmental remediation processes (lignin chelating properties), delivery of active ingredients as drug delivery vehicles, the production of natural broad-spectrum sunscreens (lignin is a UV-light absorber), and as a carrier of silver ions possessing strong antimicrobial, antifungal and antiviral activity [5, 39, 89, 90-93, 95].

Acetone had a dual role in experiments of oxidative polymerization of lignins performed by Fiţigău et al., (2013) to explore ways of enzymatic modification of lignin for high value applications [96]. First, acetone-water (50:50) (δ_H 18.8 (38.5 $MPa^{1/2}$)[4]) was used to separate a lignin fraction of lower-molecular weight (Mw <4,000 g/mol) from various lignin types including organosolv, soda, and kraft from angiosperms and softwoods. Then, the acetone-water soluble lignin fractions were oxidized by laccase, a phenoloxidase from white rot fungus *Trametes versicolor* in aqueous acetone at 50% (vol) concentration. This acetone concentration was considered optimal to maintain lignin solubility in the reaction medium for good accessibility while retaining laccase activity. A significant loss of enzyme activity was observed at acetone concentrations higher than 55%. Results showed that in these conditions lignins undergo oxidation at relatively high rates leading to a notable increase in molecular weight (M_w up to 17,500 g/mol).

Lignin in its polymer form has been studied to prepare composite matrices of hydrogels and aerogels in attempts to replace fossil-based precursors in the production of high performance functional materials. Typically, numerous synthetic polymers such as polyethyl methacrylate and polyacrylamide are used as hydrogels and resorcinol-formaldehyde and phenol-formaldehyde resins as aerogels [97, 98]. In these experiments acetone has been employed in different operations of the production sequences including washing of the polymerized product before drying during the preparation of a novel cellulose-lignin hydrogel [99] and in the solvent-exchange step that is required before drying the lignin-based aerogel in SC-CO_2 as water is not miscible with SC-CO_2. [6].

On the other side of the spectrum, lignin depolymerization is used for the production of aromatic chemicals such as benzene, toluene, xylene (BTX), and phenol. Acetone has also shown beneficial in the experiments conducted in this area. Thring et al., (2000) used acetone to dissolve organosolv (hardwood Alcell) lignin, which was then fed into a fixed bed reactor operating at 500 – 650°C for catalytic (zeolite catalyst) conversion to liquid and gaseous hydrocarbon products. They reported a complete solubility of lignin in acetone enabling direct conversion of lignin. Depending on the conditions, high yields of up to 43% of gasoline range aromatic hydrocarbons including BTX, were produced from lignin dissolved in acetone and treated with a zeolite catalyst [100]. Acetone has also been the solvent of choice in experimental depolymerization of angiosperm (hardwood and wheat straw) Alcell lignin in supercritical conditions. Lignin dissolved in acetone-water (8:2) was converted to aromatic monomers and other aromatic compounds in SC-CO_2-acetone-water-formic acid with a total yield of 10-12% based on lignin [101].

CONCLUSION

Acetone, as a designed product of the lignocellulosic biorefinery industry has demonstrated a remarkable versatility in lignocellulosic industries. Its current applications in characterization of lignocellulosics and lignin are well established. Potential applications involve a multitude of roles in the processing of lignocellulosics, spanning from delignification to produce pulp or to lessen recalcitrance to the production of high-value products from the individual lignocellulosic constituents. Acetone is distinguished for its outstanding role as an attractive solvent in studies directed to enhance lignin and promote its valorization. It may be expected that present and future lignocellulosic facilities will increasingly rely on the contribution of acetone as a biorenewable solvent, especially for the processing of lignin, the most abundant natural aromatic polymer on earth with enormous potential to reduce dependence on fossil-based aromatic products.

ACKNOWLEDGMENTS

The author would like to thank Linda Fagan for support in editing this manuscript.

REFERENCES

[1] Kamm, B. Gruber, P.R., and Kamm, M., eds. (2006): *Biorefineries-Industrial Processes and Products*, Wiley-VCH Weinheim, Germany.

[2] Hu, F. and Ragauskas, A. (2012): "Pretreatment and Lignocellulosic Chemistry," *Bioenergy Research* 5(4): 1043-1066.

[3] Zhao, X., Cheng, K., and Liu, D. (2009): "Organosolv pretreatment of lignocellulosic biomass for enzymatic hydrolysis," *Appl. Microbiol. Biotechnol.* 82: 815-827.

[4] Boeriu, C.G, Fitigău, F.I., Gosselink, R.J.A., Frissen, A., E., Stoutjesdijk, J., and Peter, F. (2014): "Fractionation of five technical lignins by selective extraction in green solvents and characterisation of isolated fractions," *Ind. Crops Prod.* 62: 481-490.

[5] Beisl, S., Miltner, A. and Friedl, A. (2017): "Lignin from micro- to nanosize: Production methods," *Int. J. Mol. Sci.* 18: 1244.

[6] Perez-Cantu, L., Liebner, F., and Smirnova, I. (2014): "Preparation of aerogels from wheat straw lignin by cross-linking with oligo(alkylene glycol)-α, ω- diglycidyl ethers," *Microporous and Mesoporous Mater.* 195: 303-310.

[7] Capello, C., Fischer, U., and Hungerbühler, K. (2007): "What is a green solvent? A comprehensive framework for the environmental assessment of solvents," *Green Chem* 9: 927-934.

[8] Alfonsi, K., Colberg, J., Dunn, P.J. et al., (2008): "Green chemistry tools to influence a medicinal chemistry and research chemistry based organisation," *Green Chem* 10:31-36.

[9] Holmbom, B. (1999): "Extractives," in *Analytical Methods in Wood Chemistry, Pulping, and Papermaking*, E. Sjöström, Alén, R., eds., Springer Series in Wood Science, T.E. Timell, ed. Springer – Verlag Germany: 125-148.

[10] Willför, S.M., Ahotupa, M.O., Hemming, J.E., et al., (2003): "Antioxidant activity of knotwood extractives and phenolic compounds of selected tree species," *J. Agric. Food Chem.* 51: 7600-7606.

[11] Pietarinen, S.P, Willför, S.M., Ahotupa, M.O., Hemming, J.E., and Holmbom, B.R. (2006): "Knotwood and bark extracts: strong antioxidants from waste materials," *J. Wood Sci.* 52: 436-444.

[12] Ralph, J., Brunow, G., and Boerjan, W. (2007): "Lignins," in: *Encyclopedia of Life Sciences.* John Wiley & Sons, Ltd. www.els.net.

[13] Sjöström, E. (1993): *Wood Chemistry: Fundamentals and Applications*, 2nd Edition, Academic Press. Inc., San Diego.

[14] Alvira, P., Tomás-Pejó, E., Ballesteros, M., and Negro, M.J. (2010): "Pretreatment technologies for an efficient bioethanol production process based in enzymatic hydrolysis: A review," *Biores. Technol.* 101:4851-4861.

[15] Amidon, T.E., Wood, C.D., Shupe, A.M., Wang, Y., Graves, M., and Liu, S. (2008): "Biorefinery: Conversion of woody biomass to

chemicals, energy, and materials," *J. Biobased Mat Bio* 2(2):100-120.

[16] Gong, C., Goundalkar, M., Bujanovic, B., and Amidon, T.E. (2012): "Evaluation of different sulfur-free delignification methods for hot-water extracted hardwood," *J. Wood Chem. Technol.* 32 (2):93-104.

[17] Fengel, D., and Wegener, G. (1984): *Wood: Chemistry, Ultrastructure, Reactions,* Walter de Gruyter, New York.

[18] Stolarski, M.J., Szczukowski, S., Tworkowski, J., and Klasa, A. (2013): "Yield, energy parameters and chemical composition of short-rotation willow biomass," *Ind. Crop. Prod.* 46: 60-65.

[19] Del Río, J.C., Renocret, J., Prinsen, P., Martinez, A.T., and Ralph, J. (2012): "Structural characterization of wheat straw lignin as revealed by analytical pyrolysis, 2D-NMR, and reductive cleavage methods," *J. Agric. Food Chem.* 60: 54922-5935.

[20] Jiménez, L., Maestre, F., and Pérez, I. (1999): "Use of butanol-water mixtures for making wheat straw pulp," *Wood Sci. Technol.* 33: 97-109.

[21] Rémond, C., Aubry, N., Crônier, D., et al., (2010): "Combination of ammonia and xylanase pretreatments: Impact on enzymatic xylan and cellulose recovery from wheat straw," *Bioresour. Technol.* 101:6712-6717.

[22] De Vrije, T., de Haas, G.G., Tan, G.B., Keijsers, E.R.P., and Claassen, P.A.M. (2002): "Pretreatment of Miscanthus for hydrogen production by *Thermotoga elfii*," *Int. J. Hyd. Ene.* 27:1381-1390.

[23] Huijgen, W.J., Reith, J.H., and den Uil, H. (2010): "Pretreatment and fractionation of wheat straw by an acetone-based organosolv process," *Ind. Eng. Chem. Res.* 49: 10132-10140.

[24] Chacha, N., Toven, K., Mtui, G., Katima, J. and Mrema, G. (2011): "Steam pretreatment of pine (*Pinus patula*) wood residue for the production of reducing sugars," *Cellulose Chem Techol.* 45 (7-8):495-501.

[25] Testova, L., Vilonen, K., Pynnönen, H., Tenkanen, M., and Sixta, H. (2009): "Isolation of hemicelluloses from birch wood: Distribution

of wood components and preliminary trials in dehydration of hemicelluloses," *Lenzinger Ber.* 87: 58-65.

[26] TAPPI (2017):T204 cm-17 "*Solvent extractives of wood and pulp,*" TAPPI Press, Technical Association of the Pulp and Paper Industry, Atlanta, GA.

[27] SCAN (2003): SCAN-CM 49:03 "*Content of acetone-soluble matter method,*" Scandinavian Pulp, Paper and Board Testing Committee, Stockholm, Sweden.

[28] Pietta, P.-G. (2000): "Flavonoids as antioxidants," *J. Nat. Prod.* 63: 1035-1042.

[29] Eklund, P.C., Långvik, O., Wärna, J.P., Salmi, T.O., Willför, S.M., and Sjöholm, R.E. (2005): "Chemical studies on antioxidant mechanism and free radical scavenging properties of lignans," *Org. Biomol. CHem.* 3: 3336-3347.

[30] Ding, F., Benaoudia, M., Bédard, P. et al., (2005): "Wood chip physical definition and measurement," *Pulp & Paper Canada* 106 (2):27-32.

[31] Boudet, A. (2000): "Lignins and lignification: Selected issues," *Plant Physiol. Biochem.* 38: 81-96.

[32] Ralph, J., Hatfield, R.D., Quideau, S. et al., (1994): "Pathway of *p*-coumaric acid incorporation into maize lignin as revealed by NMR," *J. Am. Chem. Soc.* 116:9448-9456.

[33] Ralph, J., Grabber, J.H., and Hatfield, R.D. (1995): "Lignin-ferulate cross-links in grasses: active incorporation of ferulate polysaccharide esters into ryegrass lignins," *Carbohydrate Res.* 275: 167-178.

[34] Buanafina, M.M.dO. (2009): "Feruloylation in grasses: current and future perspectives," *Molecular Plant* 2(5): 861–872.

[35] Foston, M. and Ragauskas, A.J. (2012): "Biomass characterization: Recent progress in understanding biomass recalcitrance," *Ind. Biotechnol.* 8(4):191-208.

[36] Zhang, Y.-H.P., Ding, S.-Y., Mielenz, J.R., et al., (2007): "Fractionating recalcitrant lignocellulose at modest reaction conditions," *Biotechnol. Bioeng.* 97 (2): 214-223.

[37] Vishtal, A. and Kraslawski, A. (2011): "Challenges in industrial applications of technical lignins," *BioResources* 6(3): 3547-3568.

[38] Ragauskas, A.J., Beckham, G.T., Biddy, M.J. et al., (2014): "Lignin valorization: Improving lignin processing in the biorefinery," *Science* 44: 1246843/1-10.

[39] Duval, A. and Lawoko, M. (2014): "A review of lignin-based polymeric, micro- and nano-structured materials," *React. Funct. Polym.* 85: 78-96.

[40] Lin, S.Y. and Dence, C.W., eds. (1992): *Methods in Lignin Chemistry,* Springer-Verlag Berlin Heidelberg.

[41] Björkman, A. and Person, B. (1957): "Studies on finely divided wood Part 2. The properties of lignins extracted with neutral solvents from softwoods and hardwoods," *Svensk Papperstidning* 60 (5):158-169.

[42] Nimz, H.H., Robert, D., Faix, O., and Nemr, M. (1981): "Carbon-13 NMR spectra of lignins: Structural differences between lignins in hardwoods, softwoods, grasses, and compression wood," *Holzforschung* 35: 16-26.

[43] Wu, S. and Argyropoulos, D.S. (2003): "An improved method for isolating lignin in high yield and purity," *J. Pulp Pap. Sci.* 29(7):235-240.

[44] Capanema, E.A., Balakshin, M.Y., and Kadla, J.F. (2004): "A comprehensive approach for quantitative lignin characterization by NMR spectroscopy," *J. Agric. Food Chem.* 52: 1850-1860.

[45] Capanema, E.A., Balakshin, M.Y., and Kadla, J.F. (2005): "Quantitative characterization of a hardwood milled wood lignin by nuclear magnetic resonance spectroscopy," *J. Agric. Food Chem.* 53: 1850-1860.

[46] Fujimoto, A., Matsumoto, Y., Chang, H.-M., and Meshitsuka, G. (2005): "Quantitative evaluation of milling effects on lignin structure during the isolation process of milled wood lignin," *J. Wood Sci.* 51: 89-91.

[47] Ikeda, T., Holtman, K., Kadla, J.F., Chang, H.-m., and Jameel, H. (2002): "Studies on the effect of ball milling on lignin structure using a modified DFRC method," *J. Agric. Food Chem.* 50: 129-135.

[48] Guerra, A., Fillponen, I., Lucia, L.A., and Argyropoulos, D.S. (2006): "Comparative evaluation of three lignin isolation protocols for various wood species," *J. Agric. Food. Chem.* 54: 9696-9705.

[49] Holtman, K.M., Chen, N., Chappell, M.A., Kadla, J.F., Xu, L., and Mao, J. (2010): "Chemical structure and heterogeneity differences of two lignins from Loblolly pine as investigated by advances solid state NMR spectroscopy," *J. Agric. Food Chem.* 58: 9882-9892.

[50] Goundalkar, M.J., Corbett, D.B., and Bujanovic, B.M. (2014): "Comparative Analysis of Milled Wood Lignins (MWLs) Isolated from Sugar Maple (SM) and Hot-Water Extracted Sugar Maple (ESM)," *Energies* 7(3): 1363-1375.

[51] TAPPI (2015):T222 om-15 "*Acid-insoluble lignin in wood and pulp,*" TAPPI Press, Technical Association of the Pulp and Paper Industry, Atlanta GA.

[52] Lawoko, M., Henriksson, G., and Gellerstedt, G. (2006): "Characterisation of lignin-carbohydrate complexes (LCCs) of spruce wood (*Picea abies* L.) isolated with two methods," *Holzforschung* 60: 156-161.

[53] Fasching, M., Schröder, P., Wollboldt, R.P., Weber, H.K., and Sixta, H. (2008): "A new and facile method for isolation of lignin from wood based on complete wood dissolution," *Holzforschung* 62 (1):15-23.

[54] Schuerch, C. (1952): "The solvent properties of liquids and their relation to the solubility, swelling, isolation, and fractionation of lignin," *J. Am. Chem. Soc.* 73:5061-5067.

[55] Ralph, J. and Landucci, L.L. (2010): "NMR of lignins," in *Lignin and Lignans: Advances in Chemistry*, C. Heitner, Dimmel, D., and Schmidt, J.A., eds., CRC Press Taylor & Francis Group, Boca Raton, FL: 137-244.

[56] Ralph, S.A. and Ralph, J. (2009): *"NMR database of lignin and cell wall model compounds,"* https://www.glbrc.org/databases_and_software/nmrdatabase/NMR_DataBase_2009_Complete.pdf.
[57] Li, J. and Gellerstedt, G. (2008): "Improved lignin properties and reactivity modification in the autohydrolysis process of aspen wood," *Ind. Crops Prod.* 27: 175-181.
[58] FAO (2017): "Forest Products 2015," *Food and Agriculture Organization of the United Nations,* Rome, 2017. http://www.fao.org/3/a-i7304m.pdf.
[59] Johansson, A., Aaltonen, O., and Ylinen, P. (1987): "Organosolv pulping – Methods and pulp properties," *Biomass* 13: 45-65.
[60] Oliet, M., García, J., Rodríguez, F., and Gilarranz, M.A. (2002): "Solvent effects in autocatalyzed alcohol-water pulping: Comparative study between ethanol and methanol as delignifying agents," *Chem. Eng. J.* 87(20); 157-162.
[61] Young, R.A. and Akhtar, M., eds. (1998): *Environmentally Friendly Technologies for the Pulp and Paper Industry*, John Wiley & Sons, Inc. New York.
[62] Stockburger, P. (1993): "An overview of near-commercial and commercial solvent-based pulping processes," *TAPPI J.* 76 (6): 71-74.
[63] Santucci, B.S., Bras, J., Belgacem, M.N., Curvelo, A.A.S., and Pimenta, M.T.B. (2016): "Evaluation of the effects of chemical composition and refining treatments on the properties of nanofibrillated cellulose films from sugarcane bagasse," *Ind. Crops Prod.* 91: 238-248.
[64] McDonough, T.J. (1993): "The chemistry of organosolv delignification," *Tappi J.* 76 (8): 186-193.
[65] Ding, R., Hongchao, W., Thunga, M., Bowler, N., and Kessler, M.R. (2016): "Processing and characterization of low-cost electrospun carbon fibers from organosolv lignin/polyacrylonitrile blends," *Carbon* 100:126-136.

[66] Jiménez, L., de la Torre, M.J., Bonilla, J.L., and Ferrer, J.L. (1998): "Organosolv pulping of wheat straw by use of acetone-water mixtures," *Process Biochemistry* 33(4): 401-408.

[67] Jiménez, L., García, J.C., Pérez, I., Ferrer, J.L., and Chica, A. (2001): "Influence of the operating conditions in the acetone pulping of wheat straw on the properties of the resulting paper sheets," *Biores. Technol.* 79: 23-27.

[68] Da Silva Perez, D. and Curvelo A.A.S. (2010): "Selective acetone-water delignification of *Eucalyptus urograndis*: An alternative towards the biorefinery approach," *The Open Agric. J.* 4: 145-152.

[69] Jiménez, L., Pérez, I., López, F., Ariza, J., and Rodríguez, A. (2002): "Ethanol-acetone pulping of wheat straw: Influence of the cooking and the beating of the pulps on the properties of the resulting paper sheets," *Biores. Technol.* 83: 139-143.

[70] Baeza, J., Urizar, S., Freer, J., Rodríguez, J., Pertalta-Zamora, P., and Durán, N. (1999): "Organosolv pulping. IX. Formic acid/acetone delignification of *Pinus radiata* and *Eucalyptus globulus*," *Cellulose Chem. Technol.* 33: 289-301.

[71] Zarubin, M.Y., Dejneko, I.P., Evtuguine, D.V., and Robert, A. (1989): "Delignification by oxygen in acetone-water media," *TAPPI J.* 72(11): 163-168.

[72] Pascoal-Neto, C., Delpech-Barrié, F., and Robert, A. (1993): "Délignification par l'oxygène en milieu solvent organique-eau. II. Étude comparative des bois d'eucalyptus (*E. globulus*) et de peuplier (*Populus* species)," *Cellulose Chem. Technol.* 27:185-199.

[73] Ferraz, A., Rodríguez, J., Freer, J., and Baeza, J. (2000): "Formic acid/acetone-organosolv pulping of white-rotted *Pinus radiata* softwood," *J. Chem. Technol. Biotechnol.* 75: 1190-1196.

[74] Hasan, A., Bujanovic, B., and Amidon, T. (2010): "Strength properties of kraft pulp produced from hot-water extracted wood chips within the biorefinery," *J. Biobased Mater. Bioenergy,* 4(1):46-52.

[75] Gong, C. and Bujanovic, B.M. (2014): "Impact of Hot-Water Extraction on Acetone-Water Oxygen Delignification of *Paulownia Spp.* and Lignin Recovery." *Energies* 7, no. 2: 857-873.

[76] Gong, C. and Bujanovic, B. (2016): "Effect of hot-water extraction of sugar maple on organosolv delignification and lignin recovery," *J. Bioresources Bioproducts,* 1(1):22-29.

[77] Ye, P.X., Cheng, L., Ma, H., Bujanovic, B., Goundalkar, M., and Amidon, T.E. (2012): "Biorefinery with water," in: *The Role of Green Chemistry in Biomass processing and Conversion*, H. Xie and Gathergood, N., eds. John Wiley & Sons, Inc., Hoboken, NJ: 135-180.

[78] Uygut, M., Zelie, M., Wood, C., Corbett, D., Dongre, P., and Bujanovic, B. (2015): "Potential for enhancement of enzymatic hydrolysis of sugar maple (*Acer saccharum*)," *249th ACS National Meeting*, Denver, CO. March 2015:CELL106. Poster.

[79] Zelie, M.J., Wood, C.D., Corbett, D.B., Montanaro, M.G., Bujanovic, B., Amidon, T.E. (2014): "*Processing of hot-water extracted hardwoods for the production of pulp for papermaking and enzymatic hydrolysis,*" 247th ACS National Meeting and Exhibition, March 16-19, Dallas, TX: CELL85. Poster.

[80] Paszner, L. and Chang, P.C. (1983): "*Organosolv delignification and saccharification process for lignocellulosic plant materials,*" US Patent 4,409,032.

[81] Araque, E., Parra, C., Freer, J. et al., (2008): "Evaluation of organosolv pretreatment for the conversion of Pinus radiate D. Don to ethanol," *Enzyme Microb. Technol.* 43(2): 214-219.

[82] Moxley, G., Zhu, Z., and Zhang, Y.-H.P. (2008): "Efficient sugar release by the cellulose solvent-based lignocellulose fractionation technology and enzymatic hydrolysis," *J. Agric. Food Chem.* 56: 7885-7890.

[83] Moradi, F., Amiri, H., Soleimanian-Zad, S., Ehsani, M.R., and Karimi, K. (2013): "Improvement of acetone, butanol, and ethanol production from rice straw by acid and alkaline pretreatments," *Fuel* 112: 8-13.

[84] Jones, D.T. and Woods, D.R. (1986): "Acetone-butanol fermentation revisited," *Microbiol. Rev.* 50: 484-524.

[85] Cui, C., Sun, R., and Aryropoulos, D.S. (2014): "Fractional precipitation of softwood kraft lignin: Isolation of narrow fractions common to a variety of lignins," *ACS Sustainable Chem Eng* 2: 959-968.

[86] Vangeyete, P., Gautier, S., and Jérôme, R. (2004): "About the methods of preparation of poly(ethyleneoxide)-b-poly(ε-caprolactone) nanoparticles in water Analysis by dynamic light scattering," *Colloids Surf. A: Physicochem. Eng. Aspects* 242:203-211.

[87] Ropponen, J., Räsänen, L., Rovio, S., et al., (2011): "Solvent extraction as a means of preparing homogenous lignin fractions," *Holzforschung* 65: 543-549.

[88] Hansen, C.M. and Björkman, A. (1998): "The ultrastructure of wood from a solubility parameter point of view," *Holzforschung* 52: 335-344.

[89] Frangville, C., Rutkevičius, M., Richter, A.P., Velev, O.D., Stoyanov, S.D., and Paunov, V.N. (2012): "Fabrication of environmentally biodegradable lignin nanoparticles," *Chem. Phys. Chem.* 13: 4235-4243.

[90] Lievonen, M., Valle-Delgado, J.J., Mattinen, M.-L. et al., (2016): "A simple process for lignin nanoparticle preparation," *Green Chem.* 18: 1416-1422.

[91] Richter, A.P., Bharti, B., Armstrong, H.B., et al., (2016): "Synthesis and characterization of biodegradable lignin nanoparticles with tunable surface properties," *Langmuir* 32: 6468-6477.

[92] Yearla, S.R. and Padmasree, K. (2016): "Preparation and characterisation of lignin nanoparticles; evaluation of their potential as antioxidants and UV protectants," *J. Exp. Nanosci.* 11 (4):289-302.

[93] Qian, Y. Zhong, X, Li, Y., and Qiu, X. (2017): "Fabrication of uniform lignin colloidal spheres for developing natural broad-

spectrum sunscreens with high sun protection factor," *Ind. Crops Prod.* 101: 54-60.

[94] Lu, Q., Zhu, M., Zu, Y., et al., (2012): "Comparative antioxidant activity of nanoscale lignin prepared by a supercritical antisolvent (SAS) process with non-nanoscale lignin," *Food Chem.* 135 (1): 63-67.

[95] Richter, A.P., Brown, J.S., Bharti, B., et al., (2015): "An environmentally benign antimicrobial nanoparticle based on a silver-infused lignin core," *Nat. Nanotechnol.* 10: 817-824.

[96] Fiţigău, I.F., Peter, F., and Boeriu, C.G. (2013): "Oxidative polymerization of lignins by laccase in water-acetone mixture," *Acta Biochim. Pol.* 60 (4):817-822.

[97] Grishechko, L.I., Amaral-Labat, G., Szczurek, A., Fierro, V., Kuznetsov, B.N., and Celzard, A. (2013): "Lignin-phenol-formaldehyde aerogels and cryogels," *Microporous Mesoporous Mater.* 168: 19-29.

[98] Thakur, V.K. and Thakur, M.K. (2015): "Recent advances in green hydrogels from lignin: a review," *Int. J. Biol. Macromol.* 72: 834-847.

[99] Ciolacu, D., Oprea, A.M., Anghel, N., Cazacu, G., and Cazacu, M. (2012): "New cellulose-lignin hydrogels and their application in controlled release of polyphenols," *Mater. Sci. Eng.* C. 32: 452-463.

[100] Thring, R.W., Katikaneni, S.P.R., Bakhshi, N.N. (2000): "The production of gasoline range hydrocarbons from Alcell^{R} lignin using HZSM-5 catalyst," *Fuel Process Technol.* 62:17-30.

[101] Gosselink, R.J.A., Teunissen, W., van Dam, A.E.G. et al., (2012): "Lignin depolymerization in supercritical carbon dioxide/ acetone/water fluid for the production of aromatic chemicals," *Biores. Technol.* 106: 173-177.

In: Acetone
Editors: Irenka Kozłowska et al.
ISBN: 978-1-53613-298-4

Chapter 5

ACETONE SENSOR DEVELOPMENT BASED ON TERNARY METAL OXIDES

Mohammed M. Rahman*, PhD, Mohammed M. Hussain and Abdullah M. Asiri, PhD
Chemistry department and Center of Excellence for Advanced Material Research (CEAMR), Faculty of Science, King Abdulaziz University, Jeddah, Saudi Arabia

ABSTRACT

Acetone (2-propanone or dimethyl ketone) is the simplest and most important of the aliphatic (fat-derived) ketones, which generally used as organic solvent of industrial and chemical significance. Pure acetone is a colorless, somewhat aromatic, flammable, mobile liquid that boils at 56.2°C (133°F). Construction of highly efficient toxic chemical sensors is the key approach for the determination of carcinogenic chemicals in the environment and ecosystem. We report here, an efficient acetone chemical sensor based on the analytical performances such as sensitivity, limit of detection (LOD), good linearity, and reproducibility. The

* Corresponding Author Email: mmrahman@kau.edu.sa; Phone: +00966596421830.

proposed acetone-detecting electrode was introduced by the implementation of ternary metal oxides nanoparticles (TMO NPs) as a successful electron mediator with glassy carbon electrode (GCE) assembly. The prepared TMO NPs were well crystalline-doped nanomaterial (NMs) and produced by implementation of hydrothermal procedure at low temperature. The conventional methods such as Fourier-transform infrared spectroscopy (FTIR), ultraviolet visible spectroscopy (UV-vis), field emission scanning electron microscopy (FESEM), X-ray photoelectron spectroscopy (XPS), energy-dispersive X-ray spectroscopy (EDS), and powder x-ray diffraction (XRD) were utilized for characterization of prepared NPs in terms of morphological, elemental, optical and structural properties. The linear dynamic range (LDR) of 0.34 $nmolL^{-1}$ to 3.4 $mmolL^{-1}$ with lower LOD (S/N = 3) and a higher sensitivity were exhibited by lab-made fabricated sensor based on TMO NPs for selective acetone detection. In shortly, the TMO NPs are utilized as an excellent electron mediator with Nafion/GCE assembly in a chemical sensor for acetone detection even at the very low concentration. Therefore, the chemical sensor is fabricated with TMO NPs may be a promising highly sensitive sensor by reliable I-V detection method for the effective detection of hazardous and carcinogenic chemicals in medical as well as health-care fields.

INTRODUCTION

Generally, acetone is the simplest and most significant of the ketones, which is a colorless, mobile, flammable liquid with a mildly pungent, somewhat aromatic odor, and is miscible in all proportions with water and the most organic solvents. Pure acetone is a colorless, somewhat aromatic, flammable, mobile liquid that boils at 56.2°C (133°F). It is an excellent solvent for a wide range of gums, waxes, resins, fats, greases, oils, dyestuffs, and cellulosic. It is used as a carrier for acetylene for the manufacturing of a variety of coatings and plastics. This is used as a raw material for the chemical synthesis of a wide range of products such as ketene, methyl methacrylate, bisphenol A, diacetone alcohol, methyl isobutyl ketone, hexylene glycol, methyl, pentanediol, and isophorone etc. The acetone has diversified industrial application in multiple areas and since it is highly volatile chemical, it may cause headache, fatigue in human due to breathing of acetone even at ppm level [1]. Therefore, the analysis of breath is a key tool in the public health sector to investigate a large range of diseases which becoming more popular from the last five years [2-4]. The third major reason for death is diabetes in the industrialized countries [5], and according to breath diagnosis report, maximum 0.8 mgL^{-1} acetone concentration is in normal health human, while higher than 1.8 mgL^{-1} is in diabetic patient [6]. Beside this, the central nervous system of human body may be affected seriously, if the acetone concentration is more than normal in blood [7]. Therefore, it is urgent needed to develop of highly sensitive and reliable method to detect acetone in environment, work please and as well as diagnosis the human breath [8, 9].

Acetone was a coproduct of the Shell process for glycerol. Propylene was hydrated to isopropyl-alcohol. Some of the alcohol was catalytically oxidized to acrolein and some was oxided to give hydrogen peroxide and acetone. Some more of the isopropyl alcohol and the acrolein reacted to give allylalcohol and acetone. The allyl alcohol was then treated with the peroxide to give glycerol. About half kilogram of acetone resulted per kilogram of glycerol. Direct oxidation of hydrocarbons and catalytic oxidation of isopropyl alcohol have also been used for commercial production of acetone. Most of the world's acetone is now obtained as a coproduct of phenol by the cumene process. About 70% acetone is produced by this process. Cumene is oxidized to the hydro-peroxide and cleaved to acetone and phenol. The yield of acetone is believed to average about 4%, unit weight of acetone is obtained per unit of phenol. Moreover, to monitoring the industrial process, environment and toxin gases storages, the sensors and actuators are becoming the most important area of research [10].

LITERATURE SURVEY

The undoped or doped metal oxides of semiconductor have been implemented as a successive and efficient sensor to identify harmful and toxic in the diverse areas [11] and the most conventional sensor materials like as SnO_2 [12, 13], ZnO [14, 15], In_2O_3 [16, 17], Fe_2O_3 [18], Co_3O_4 [19], TiO_2 [20], WO_3 [21], have been already investigated as toxin gas sensor. However, most of them, ZnO with band-gap energy of 3.3 eV, have been comprehensively considered as acetone sensing material and it has various advantages such as low manufacture cost, high electron mobility and good thermal and chemical stability [22]. Another talented semiconductor metal oxide is SnO_2, which was experimented as acetone sensor, but it is limited by slow response time with low sensitivity [23]. According to S. B. Khan et al., ZnO NPs based acetone sensor was showed a better sensitivity and lower LOD with short response time of 10 sec [24]. Another report claimed that, SnO_2/ZnO hetero nano-fiber was exhibited high performance for acetone gas sensing [25].

Acetone is capable of dissolving many fats and resins as well as cellulose ethers, cellulose acetate, nitrocellulose, and other cellulose esters. Because of the latter quality, acetone is used extensively in the manufacture of artificial fibers (such as some rayon's) and explosives. It is used as a chemical intermediate in pharmaceuticals and as a solvent for vinyl and acrylic resins, lacquers, alkyd paints, inks, cosmetics (such as nail-polish remover), and varnishes. It is used in the preparation of paper coatings, adhesives, and heat-seal coatings and is also employed as a starting material in the synthesis of many compounds. Here, due to very high toxicity of acetone, it is a serious risk of illness for the public health, and it should be taken initiative for significant detection of acetone by conventional technique with TMO NPs and GCE electrode assembly system. The exploration of acetone using of TMO NPs, which was placed on head of GCE as thin films with conducting binder and the details studies were performed to construct of acetone sensing device to protect public health sector as well as the environment. The thin layer of TMO NPs with accompanying conducting binder agent was applied onto GCE to fabricate the working electrode and this electrode was performed to detect the selective analytes (acetone) applying the electrochemical approaches current vs. potential (I-V) system. It should be mentioned that the proposed acetone chemical sensor is exceptional and decent invention with active TMO NPs onto GCE.

METHODOLOGY

Materials and Methods

The chemical reagents (laboratory grade) like as ammonium hydroxide, nafion (5% ethanolic solution), acetone, 2-nitrophenol, 3-methoxyphenol, 4-aminophenol, 4-methoxyphenol, bisphenol A, ethanol, hydrazine, methanol, p-nitro-phenol, monosodium phosphate and disodium phosphate were bought from marketing company of the Sigma-Aldrich, and any additional refinement was not applied before use. The modern and highly reliable equipment, namely such as thermo scientific NICOLET iS50 FTIR spectrometer (Madison, WI, USA) and 300 UV-Vis spectrophotometer (Thermo scientific) were implemented to record FTIR and UV-Vis spectra of TMO NPs. To quantify the binding energies (eV) among Zn, Sn, Yb and O, the XPS investigation was executed for TMO NPs on a K-α1 spectrometer (Thermo scientific, K-α1 1066) with an excitation radiation source (A1 Kα1, Beam spot size = 300.0 μm, pass energy = 200.0 eV, pressure~ 10^{-8} torr). The optical characterization of synthesized NPs of $ZnO/SnO_2/Yb_2O_3$such as the arrangement of molecule, analysis of element, morphological structure, and particle size, were examined by the implementation of FESEM (JEOL, JSM-7600F, Japan) analysis and this machine is also equipped XEDS. To identify the crystallinity of NPs material, the synthesized TMO NPs were investigated by the applying of XRD assessment at atmospheric temperature. The current vs. potential (I-V) technique was accomplished to detect acetone at the desire range of electrical potential by Keithley electrometer (6517A, USA) and the fabricated TMO NPs (coating with nafion) assembly was functioned as working electrode in desire responsive buffer system.

Preparation of TMO NPs by Hydrothermal Process

Zinc chloride ($ZnCl_2$), ytterbium chloride ($YbCl_3$), tin(IV) chloride ($SnCl_4$) and ammonium hydroxide (NH_4OH) were used as reacting precursors to the preparation of TMO NPs by hydrothermal technique. The hydrothermal technique is a usual system, which is extensively applied to the synthesis of doped NMs, and the resultant achieved NMs are smaller with grains size in the reasonable smaller of phase formation. According to this technique, a conical flask (250.0 mL) was filled-up with 100.0 mL distilled water, and then $ZnCl_2$, $YbCl_3$, and $SnCl_4$ were dissolved in this conical flask, and kept it with continuous magnetic stirring. The NH_4OH was added with the produced solution to fix the pH value of the solution at 10.5, and the resultant solution was kept in the oven at 150ºC with continuous magnetic stirring. Finally, the produced NPs were placed in the muffle furnace for calcinations at 500ºC around 6 h. Then the calcined sample was characterized using FTIR, XRD, UV, XPS, and FESEM.

$$NH_4OH_{(s)} \rightarrow NH_4^+{}_{(aq)} + OH^-{}_{(aq)} \quad \text{(i)}$$

$$ZnCl_{2(s)} \rightarrow Zn^{2+}{}_{(aq)} + 2Cl^-{}_{(aq)} \quad \text{(ii)}$$

$$YbCl_{3(s)} \rightarrow Yb^{3+}{}_{(aq)} + 3Cl^-{}_{(aq)} \quad \text{(iii)}$$

$$SnCl_{4(s)} \rightarrow Sn^{4+}{}_{(aq)} + 4Cl^-{}_{(aq)} \quad \text{(iv)}$$

$$NH_4^+{}_{(aq)} + 4OH^-{}_{(aq)} + Yb^{3+}{}_{(aq)} + Sn^{4+}{}_{(aq)} + Zn^{2+}{}_{(aq)} + Cl^-{}_{(aq)} \rightarrow Yb(OH)_{3(aq)}\downarrow + Sn(OH)_{4(aq)}\downarrow + Zn(OH)_{2(aq)}\downarrow + NH_4Cl_{(aq)} \quad \text{(v)}$$

$$2Yb(OH)_{3(aq)}\downarrow + Sn(OH)_{4(aq)}\downarrow + Zn(OH)_{2(aq)}\downarrow \rightarrow Yb_2O_3\text{-}ZnO\text{-}SnO_{2(s)}\downarrow \text{(TMO)} + 6H_2O_{(aq)} \quad \text{(vi)}$$

Accordingly, the equation (i) to (vi), the proposed reactions are advanced gently and during the formation of TMO NPs in the reaction medium, the pH plays an significant accountability. When the pH of the reaction medium is adjusted at 10.5 by adding ammonium hydroxide, the $ZnCl_2$ is started to hydrolysis and converted to zinc hydroxide [equation (v)]. Therefore, the NH_4OH is the responsible for supply of OH^- and as well, as the creation of alkaline reaction medium. At the time being, Zn^{2+} and OH^- ions are increased and touched the critical value of its, and $Zn(OH)_2$ is started to form nuclei. On the basis of reaction (iii) and (v), with the increase of Yb^{3+} ions in the reaction medium, the crystal formation of $Zn(OH)_2$ becomes gentler, because of the lower activation energy of $Zn(OH)_2$ to form heterogeneous nucleation. Due to the higher existence of Yb^{3+}andSn^{4+}ions in the reaction medium, $Zn(OH)_2.Sn(OH)_4.Yb(OH)_2$ crystals are started to aggregate and formed NPs. After that, the resultants NPs are washed with water, ethanol and acetone successively. At the end, the synthesized NPs are dried overnight at the room temperature, and the calcinations of prepared NPs are carried out in the furnace (Barnstead Thermolyne, 6000 Furnace, USA) at 500 °C around 6 h. Accordingly, to Ostwald-ripening method for NPs growth mechanism, at the very beginning, TMO NPs nucleus are progressed by self and mutual-aggregation and again re-aggregated to form TMO NPs. The inside molecular arrangements of nanocrystal are with each other counterpart through Vander-Waals forces. The growth patterns of TMO NPs are similar with the previously published reports [26-30]. The formation mechanism of TMO NPs is presented in Scheme 1. This is the first time, the produced calcined TMO NPs are applied to the selective detection of acetone and regarding this, there is not any report available.

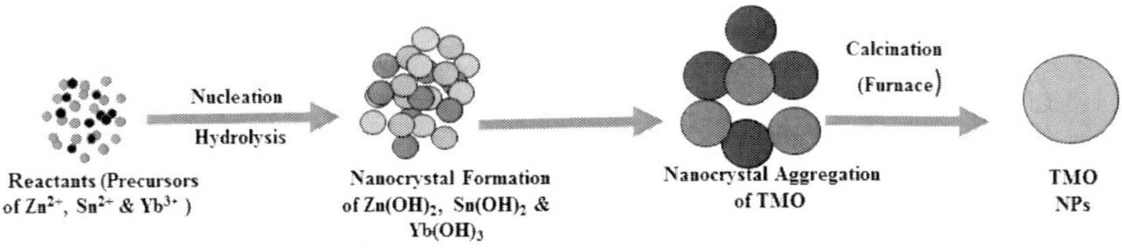

Scheme 1. The growth mechanism of TMO NPs produced by low temperature hydrothermal method.

Fabrication of GCE with TMO NPs

0.1 M phosphate buffer (PBS-solution) at pH 7.0 was prepared by mixing of equi-molar concentration of 0.2 M Na_2HPO_4 and 0.2 M NaH_2PO_4 solution in 100 mL de-ionized water at room conditions. To fabricate the working electrode with GCE, slurry of TMO NPs was prepared with ethanol, coated onto GCE and dried at the room conditions. For the improvement of the adhesive property between GCE and TMO NPs, a drop of 5% ethanolic solution as conducting coating binder was added on fabricated working electrode. Then the modified electrode was placed inside an oven at 35°C temperature and kept it until the conducting film was dried entirely. An electrochemical cell was combined with TMO NPs/GCE and Pt-wire (dia: 1.5 mm) where TMO NPs/GCE was acted as working electrode and Pt-wire was counter electrodes respectively. The thickness of the fabricated electrode was calculated as 0.1 mm. A range of acetone solution (full concentration range: 0.034 $nmolL^{-1}$ ~ 0.34 $molL^{-1}$) was prepared and implemented in the assembled electrochemical cell as the targeted analyte. In the acetone detection process, the sensitivity of the prepared working electrode based on TMO NPs was measured from the slop of the calibration curve (The ratio of current versus concentration). In the same way, the LOD, and LDR were calculated from the ratio of 3N/S. The utilized Electrometer for current vs. potential (I-V) measurement, is simple two electrode system. Amount of 0.1 M PBS solution was kept constant in the beaker as 10.0 mL throughout the chemical investigation.

Optical and Structural Analyses

The photosensitive is the major measurable characteristic for the investigation of photo-catalytic activities of prepared TMO NPs. Therefore, the synthesized NPs were investigated by UV-vis spectrum. Due to the scanning with the visible light source, the outer electron of the atom absorbs the radian energy at desire wavelength and it is transited from lower to higher energy level to result a UV-Vis spectrum [31], which is illustrated in Figure 1(b). The UV-vis spectroscopy is performed in the range of 200 nm to 800 nm at room temperature. The observed spectrum shows an intense and wider spectrum at 307 nm and it is only identical band, which is accredited for the electronic transition of the valence band of TMO NPs from lower to higher energy level. The obtained UV-Vis spectrum allows us to estimate the optical band gap (E_{bg}) and according to equation (vii), the calculated band gap energy is 4.04 eV. Here, E_{bg} = band gape energy, λ_{max} = maximum absorbed wave length [32-36].

$$E_{bg}=1240/\lambda_{max} \quad \text{(vii)}$$

The FTIR is the modern analyzing tool, which can determine the functional properties of the atom or molecule from corresponding the atomic and molecular vibration. TMO NPs were implemented to FTIR investigation, and the FTIR is performed at range of 400–4000 cm^{-1} atmospheric condition, which is illustrated in Figure 1(a). From the FTIR, several identical peaks at 532, 1626, and 3399 cm^{-1} are observed and the existence of sharpest and major peak at 436 cm^{-1} is related with metallic (Zn-O) oxygen bonds [37]. The peaks 1626 and 3399 cm^{-1} are represented the adsorbed water vibrating and O-H stretching mode, separately [38].

The phase identification is the major concern to nano-researcher and it can convey the details evidence about a unit cell dimension. To investigation of structural properties and crystal phase of synthesized TMO NPs, the XRD (powder X-ray diffraction) was implemented. The XRD machine is equipped with the radiation source Cu-K1 (1.54 A°) with the range of 10–80° and scanning speed is 2°/min. In Figure 1(c), the peaks are demonstrated in XRD pattern, the synthesized TMO NPs are consisting of well-assorted phases of ZnO, $SnO_2$and Yb_2O_3. The dedicated reflected XRD peaks for ZnO indices (μ) such as (101), (103), (110), (200), and (220) are analogous with previously reported of bulk ZnO [39, 40]. Besides of this, the observed reports are in great similarities through the typical JCPDS card no. 36-1451. In addition, the several peaks of SnO_2 indices (θ) are also noticeable into the XRD diffraction pattern, which are identified as (101), (110) and (202). The observed peaks of SnO_2 are well matched with the literature [41, 42] and JCPDS No. 41-1445. Moreover, the other peaks of Yb_2O_3indices (β) are also identified in the XRD such as (102), (110) and (202) and well matched with the literatures [43]. There is a provision to measure the crystal size of NMs from XRD diffraction pattern by the implementation of Scherer equation, which is presented in equation (viii). Here, λ is representing the wavelength (X-ray radiation = 1.5418 Å) and β is width at half, corresponding to highest intense peak, and θ is the diffracting angle. According to the Scherer equation, the calculated crystal size is 6.2 nm [44-46].

$$D = 0.9\lambda/(\beta\cos\theta) \quad \text{(viii)}$$

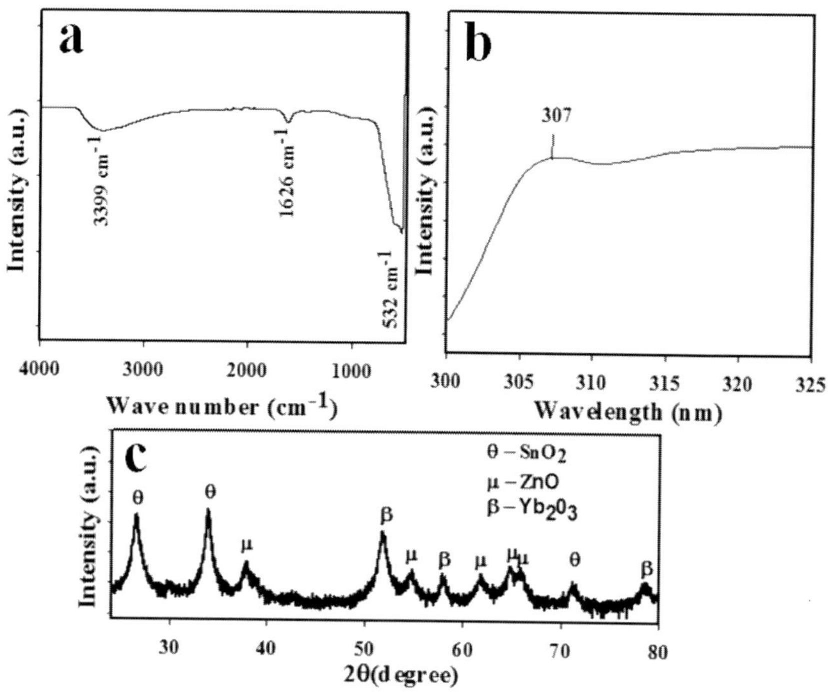

Figure 1. Optical and morphological evaluation by (a) FTIR, (b) UV spectrum, and (c) XRD pattern of the TMO NPs.

Morphological and Elemental Analyses

The FESEM investigation was carried out for the morphological studies of synthesized TMO NPs and the resultant representative low and high magnifying images of FESEM are presented in Figure 2(a-c). From the SEM image, the diameter (size) of the NPs is calculated as 13.3 nm, in the range of 6.0 to 16.0 nm. From the FESEM investigation report, it is confirmed that the produced nano-doped materials are NPs in shape [47], which is similar with the compositions of EDS analysis illustrated Figure 2(d-e).The composition of prepared NPs are authenticated from the EDS investigation. Therefore, based on the EDS analysis, Zinc (Zn), Oxygen (O), Tin(Sn) and Ytterbium (Yb) are existed in NPs and contained O 32.48 wt%, Zn 6.45 wt%, Sn 45.85 wt% and Yb15.22 wt%. No other supplementary peaks are detected and the conformation about the components of O, Zn, Sn and Yb existence in the prepared NMs are accredited by EDS analysis [48-50].

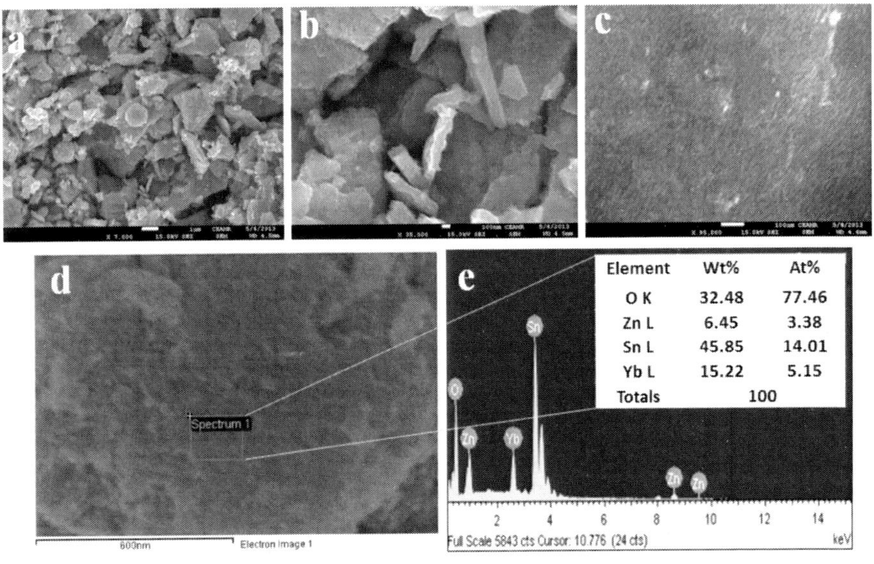

Element	Wt%	At%
O K	32.48	77.46
Zn L	6.45	3.38
Sn L	45.85	14.01
Yb L	15.22	5.15
Totals	100	

Figure 2. (a-c) FESEM analysis from low to high magnified images and (d-e) elemental analysis of TMO NPs.

Binding Energy Analysis

The quantitative spectroscopic technique XPS was applied to investigate the nature of species existing in prepared TMO NPs. Due to the kinetic energy of X-ray beam of XPS, the electrons are existed to result XPS spectrum. Generally, the atomic composition with corresponding chemical formula, the oxidation state of the species present in the NMs can be scrutinized efficiently using this practice [51]. The XPS spectrum is shown in Figure 3, which is established the existence of O, Zn, Sn and Yb although signals created from synthesized NPs and possible impurities are not detected. The main two identical peaks of Zn_{2p} orbit, there is not any apparently change between these two peaks and the both peaks are the indication of the equal oxidation state of Zn, in the sample.

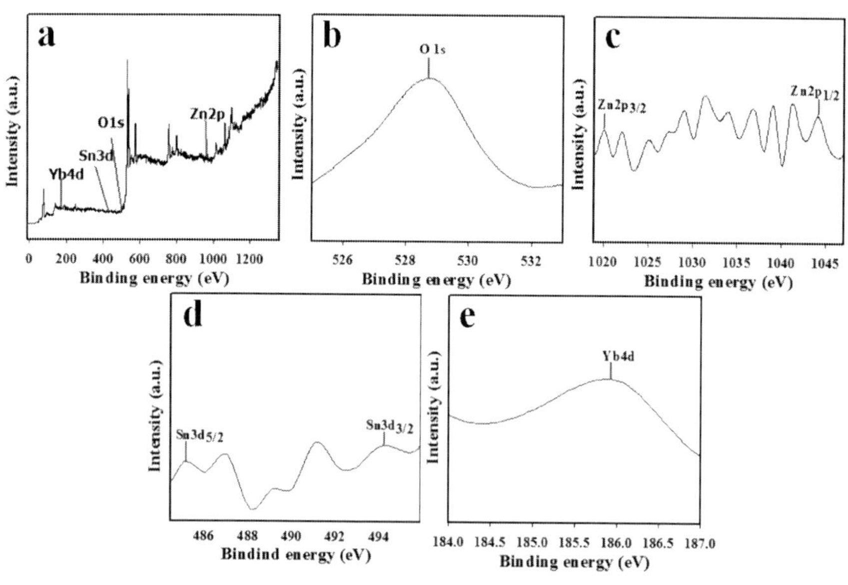

Figure 3. Binding energy analysis by XPS study of TMO NPs (a) full spectrum (b) O1s (c) spin orbit Zn2p level (d) spin orbit Sn3d level and (e) spin orbit Yb4d level.

In Figure 3(c), the observed binding energy of $Zn2p_{3/2}$ and $Zn2p_{1/2}$ peaks are 1020 and 1044 eV correspondingly, which are the characteristic value of the oxidation of Zn^{2+}, in ZnO [52]. The O 1s spectra of the ZnO sample, the peak is clearly recognized and centered at 529 eV, which is accredited to the lattice oxygen of ZnO [53], showed in Figure 3(b). The Sn XPS spectrum, which is shown in Figure 3(d), the characteristic two peaks of $Sn3d_{5/2}$ and $Sn3d_{3/2}$ are identified and centered at 485 eV and 494 eV respectively. The spectra is significantly conformed the attribution of Sn^{4+} state [54, 55]. The other spectrum of Yb_{4d} is visible and centreted at 186 eV presented in Figure 3(e) [56].

APPLICATIONS

Detection of Acetone by TMO NPs

The possible use of acetone sensor constructed based on TMO NPs/GCE assembly was performed to detect and quantity the desire toxin (acetone) in buffer system. Development of acetone sensor fabricated based on the TMO NPs/GCE assembly is in the first stage and regarding this, any other article is not found. The proposed chemical sensor has several benefits such as highly stable in air, inert in chemical environment, enhanced electrochemical activity during the detection, easy to perform, assemble and fabrication, safe chemo-characteristic. The probable application of chemical sensor with TMO NPs coated onto GCE with conducting binder was evaluated in details for the detection of targeted analyte (acetone) in a suitable buffer system. The TMO NPs were acted as brilliant electron mediator during the sensing performance. To prepare a chemical sensor selective for acetone, TMO NPs were mixed with ethanol to form slurry, which was used to coat on the GCE and the assembled electrode was dried at ambient temperature. Lastly, to improve the adhesive (conducting binding properties) properties of the NPs on GCE, a drop of nafion was added, and again dried systematically. The organized TMO NPs/GCE assembly was performed as a working electrode to construction of the chemical sensor, where the targeted analyte was acetone. During the sensing operation of acetone in desire buffer system, the applied current vs. potential (I-V) was assessed on thin-film of TMO

NPs/GCE in the working electrode, which was increased noticeably due to the adsorption of aqueous acetone on the fabricated working electrode. 1.0s was set as holding period in the electrometer and an amplified current vs. potential was observed during acetone detection by fabricated chemical sensor. A proposed oxidation mechanism of acetone is illustrated the in Scheme 2. Due to the oxidation of acetone, an enrichment of electron in the buffer medium is observed and it is caused amendment of I-V response. Based on the proposed acetone oxidation mechanism, the O_2 and acetone are adsorbed on surface of working electrode and immediately the acetone is oxidized to produce electron, which increases the potential of the measuring buffer system. The pictorial application of the TMO NPs modified GCE electrode for acetone chemical sensor is demonstrated in Scheme 2.

The constructed working electrode of acetone sensor based on TMO NPs was applied to detection of acetone in desire buffer medium and the proposed oxidation mechanisms of acetone is presented in reaction (ix) to (x). According to the electrochemical oxidation process of acetone, the several electrons are released, which is enhanced the I-V response of TMO NPs/GCE sensor.

$$CH_3COCH_3 + 4O^- \rightarrow CH_3COOH + CO_2 + H_2O + 4e^- \quad \text{(ix)}$$

$$CH_3COOH + 7O^- \rightarrow 3CO_2 + 3H_2 + O_2 + 7e^- \quad \text{(x)}$$

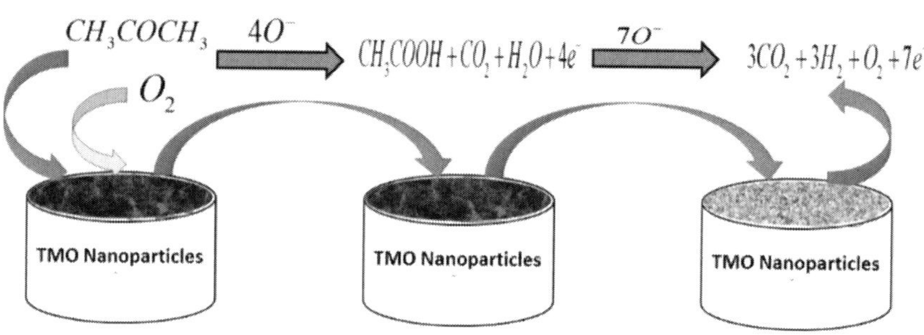

Scheme 2. The expected mechanism for determination of acetone by TMO NPs by electrochemical approach.

To optimize the pH, the fabricated electrode was applied in various buffer system (pH = 5.7, 6.5, 7.0, 7.5 and 8.0) and it was found more responsive at pH 7.5, which is presented in Figure 4(a). During this performance, by using the Keithley electrometer, the applied electrical potential range was 0 to + 1.5 V. The several chemical (toxin) reagents such as acetone, 2-nitrophenol, 3-methoxyphenol, 4-aminophenol, 4-methoxyphenol, bisphenol A, methanol, ammonium hydroxide, and p-nitro-phenol were applied to study the selectivity of fabricated chemical sensor. Based on selectivity study, acetone is exhibited the supreme current response toward the TMO NPs/GCE electrode which is illustrated in Figure 4(b).

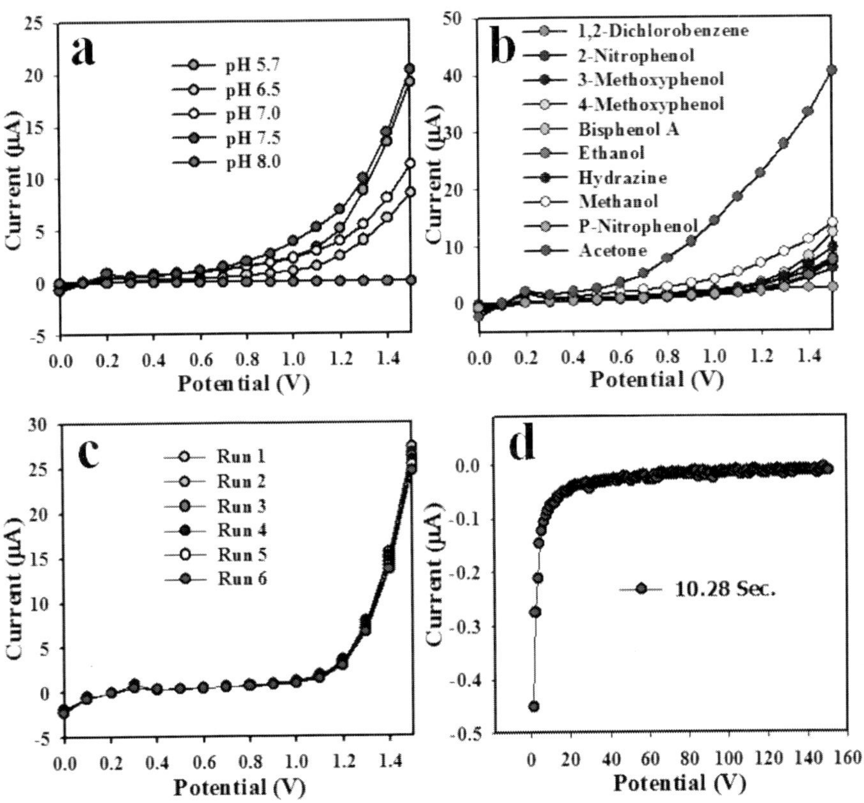

Figure 4. Optimization of acetone sensor with TMO NPs, (a) pH optimization, (b) selectivity, (c) repeatability, and (d) response time.

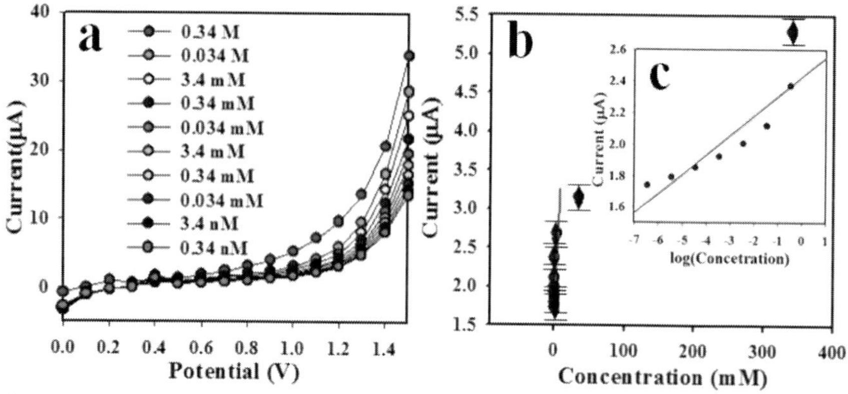

Figure 5. (a) Concentration variation of acetone sensor based on TMO NPs/GCE by I-V method and (b) calibration curve (Inset: log [acetone conc.] vs current).

To find out the reproducibility of TMO NPs coated GCE, the sensing current was measured. It was measured the seven measurements of acetone solution in 0.34 $\mu molL^{-1}$ concentration which is presented in Figure 4(c). The outstanding reproducible responses are found with invented electrode in the indistinguishable surroundings and it is declared that the current vs. potential (I-V) responses are unaltered expansively even after washing of working electrode in each trial. The standard deviation of reproducibility (RSD) of invented GCE is measured and it is 2.74%. The response time is another tool for the measurement of efficiency of the fabricated TMO NPs/GCE; the test was performed in 0.34 $\mu molL^{-1}$ acetone solution and the calculated response timeis10.28 seconds, which is illustrated in Figure 4(d).

The fluctuating in the current response of working electrode is directly proportional to the concentration of acetone in desire buffer medium under standard condition. Therefore, the I-V responses of the modified TMO NPs/GCE sensor system were measured at the different concentration of acetone. It is concluded that the current vs. potential (I-V) response is amplified with enrichment of the concentration of the electrolyte (acetone) in the buffer system. A sequence of analyte concentration (0.34 $nmolL^{-1}$ - 0.34 $molL^{-1}$) were estimated from a lower to higher potential (0 to +1.5V) showing in Figure 5(a). To scale up analytical limit of the sensor, a linear calibration curve were plotted at potential +1.0 V from the diverse (lower-higher) concentration of acetone solution shown in Figure 5(b) and the regression coefficient (r^2 = 0.9394), sensitivity (S = 17.0886 μA.mmol.$L^{-1}cm^{-2}$), a LDR = 0.34 nmol.L^{-1} to 3.4 mmolL^{-1} and a LOD = 0.05 ± 0.002 nmolL^{-1}) were calculated from the curve of calibration where signal to noise ratio was 3.

In the electrochemical sensing system, the current *vs* applied potential is directly proportional to the corresponding electrolyte (acetone) concentration in buffer medium and an amplified current signal was visible when the acetone concentration was increased in the measuring system. The similar tendency of current vs potential with concentration of toxin analyses were reported in the several literatures [57, 58]. In desire measuring buffer system, the targeted toxin (acetone) molecules are occupied a small number of surface coverage on TMO NPs/GCE film and the oxidation reaction of the acetone is started progressively. Therefore with the increasing of acetone concentration in buffer medium, the reaction on the anticipated working electrode surface is increased remarkable and the larger surface coverage with the corresponding acetone molecules are obtained. Again additional enrichment of the acetone concentration in the detecting buffer medium, a highly moderate current vs potential (I-V) response is occurred and the surface coverage by the corresponding analyte on the surface of TMO NPs/GCE working electrode is maximum. Due to the frequently enhancement of analyte concentration in the medium, the surface coverage by acetone molecules is touched to the saturation. Therefore, it can be summarized that the proposed acetone chemical sensor based on TMO NPs/GCE assembly can be engaged for successive detection and quantification of the targeted toxin (acetone). As we observed, the proposed acetone sensor is showed very short response time at around 10 sec and it should be mentioned that the 10 s is needed by the desire acetone chemical sensor fabricated based on TMO NPs/GCE to become steady state saturation. Since, the indicated acetone chemical sensor exhibited moderately high sensitivity of 17.09 $\mu AmmolL^{-1}cm^{-2}$, so it can be accredited that it has supper adsorption capacity, active catalytic decomposition ability, and excellent biocompatibility [59-61]. In term of higher sensitivity and lower LOD, the proposed acetone chemical sensor based on TMO NPs is showed a reasonable higher performance compare to earlier report chemical sensor fabricated based the diversified transition nanostructured metal oxides and a comparison is illustrated in Table 1. With the high surface area and crystalline TMO NPs, the proposed electron mediator can facilitated a satisfactory nano-atmosphere which can provide the efficient acetone detection and quantification in desire buffer system. Due to the higher sensitivity of tested acetone chemical sensor can affords highly electron penetration atmosphere, which can face lifted to transfer the electrons from the active sites of TMO NPs to GCE. Therefore it can be mention that the demonstrated acetone chemical sensor, is simple and sophisticated to detection and quantification of acetone by applying current vs potential (I-V) electro-chemical approaches. The sensing performance of earlier testified metal oxides based acetone sensors are associated and summarized in Table 1. The based on performances such as LOD, LDR and sensitivity, the demonstrated acetone chemical sensor showed the better enactment.

The Analysis of Real Sample

For the validation, the fabricated acetone sensor based on TMO NPs coated GCE was applied to detect the selective chemical in real samples with various concentration of acetone. A standard addition method applied to measure the concentration of acetone in real samples [67-69]. We were used to measure the acetone in different environmental samples such as industrial effluent, PC-baby-bottle, PC-drinking-water-bottle, and PVC-food-packaging-bag. The results are showed us that the acetone detection is possible in environmental samples, which is showed in Table 2.

Table 1. Comparison of acetone sensor performances with various fabricated materials

Modified Electrode	Methods	LOD	LDR	pH	Sensitivity	Ref
ZnO nanoparticles	I–V	1.18 $\mu molL^{-1}$	0.13 $mmolL^{-1}$~ 0.13 $molL^{-1}$	-----	0.14 $\mu AmmolL^{-1}cm^{-2}$	[62]
ZnO/Co_3O_4 nanorods	I-V	14.7 $\mu molL^{-1}$	66.8 $\mu molL^{-1}$ ~ 0.133 $molL^{-1}$	7.0	3.58 $\mu AmmolL^{-1}cm^{-2}$	[63]
Ag_2O microflowers	I–V	0.11 $\mu molL^{-1}$	0.13 $\mu molL^{-1}$ ~ 0.67 $molL^{-1}$	-----	1.60 $\mu AmmolL^{-1}cm^{-2}$ mM^{-1}	[64]
Lead foil electrode	AM	50.0 mgL^{-1}	50.0 ~250.0 mgL^{-1}	-----	2.07 $\mu AmmolL^{-1}$ cm^{-2}	[65]
Mn/ZnS QDs	PA	0.2 mgL^{-1}	1.0 ~ 600.0 mgL^{-1}	12.0	---	[66]

LOD = Limit of detection, AM: Amperometry, PA: Phosphorescence attenuation, and LDR = Linear dynamic range.

Table 2. Measured concentration of the acetone in real samples

Real Sample	Observed current (μA)				Average	Calculated Conc.(μM)	SD
	R1	R2	R3	R4			
Industrial effluents	0.10	0.95	0.91	0.89	0.94	0.17	0.05
PC baby bottle	0.63	0.61	0.59	0.58	0.60	0.11	0.02
Safa water bottle	1.48	1.43	1.42	1.43	1.44	0.25	0.03
PVC food packaging bag	0.90	0.82	0.85	0.81	0.84	0.15	0.03

*R = reading, and SD = standard deviation.

CONCLUSION

In this approach, acetone sensor was developed by used mixed various metal oxides. Here, the ternary TMO NPs were produced by hydrothermal process, which characterized of the NPs were done by conventional techniques such as FESEM, EDS, FTIR, UV-Vis, XPS, and XRD. The selective acetone chemical sensor based on TMO NPs coated onto GCE with conducting binder was exhibited as a highly efficient electron mediator during the oxidation of acetone in PBS. Therefore, the fabricated working electrode is showed the higher potential and outstanding electro-catalytic activities to the acetone at a reduced over potential. The fabricated acetone chemical sensor has lower LOD with the higher sensitivity in broad LDR for acetone detection. It can be used for the detection of selective chemical in the existence of toxic and hazardous chemical in the environment for the safety of healthcare field.

REFERENCES

[1] L. K. Bagal, J. Y. Patil, I. S. Mulla, S. S. Suryavanshi, Studies on the resistive response of nickel and cerium doped SnO_2 thick films to acetone vapor, *Ceramics Inter*. 38 (2012) 6171–6179.

[2] M. Righettoni, A. Tricoli, S. Gass, A. Schmid, A. Amann, S. E. Pratsinis, Breath acetone monitoring by portable Si: WO_3 gas sensors, *Anal. Chim. Acta*. 738 (2012) 69-75.

[3] M. Righettoni, A. Tricoli, S. E. Pratsinis, Si:WO_3 sensors for highly selective detection of acetone for easy diagnosis of diabetes by breathe analysis, *Anal. Chem*. 82 (2010) 3581-3587.

[4] D. H. Kim, Y. S. Shim, J. M. Jeon, H. Y. Jeong, S. S. Park, Y. W. Kim, Vertically ordered hematite nanotube array as an ultrasensitive and rapid response acetone sensor, *ACS Appl. Mater. Interfaces* 6 (2014) 14779-14784.

[5] S. Salehi, E. Nikan, A. A. Khodadadi, Y. Mortazavi, Highly sensitive carbon nanotubes–SnO_2 nanocomposite sensor foracetone detection in diabetes mellitus breath Sara, *Sens. Actuators B* 205 (2014) 261–267.

[6] S. Wei, J. Zhao, W. Du, Synthesis, characterization and acetone-sensing properties of bristle grass-like ZnO nanostructure, *Ceramics Intl.* 41 (2015)769–776.

[7] F. M. Li, X. B. Li, S. Y. Ma, L. Chen, W.Q. Li, C.T. Zhu, X. L. Xu, Y. Chen, Y. F. Li, G. Lawson, Influence of Ce doping on microstructure of ZnO nanoparticles and their acetone sensing properties, *J. Alloys compd.* 649 (2015) 1136-1144.

[8] M. Yan, L. H. Huo, H. Zhao, S. Gao, Z. M. Rong, Electrical properties and acetone sensing characteristics of $LaNi_{1-x}TixO_3$ perovskite system prepared by amorphous cetrate decomposition, *Sens. Actuators B* 143 (2009) 111–118.

[9] P. Song, Q. Wang, Z. Yang, Preparation, characterization and acetone sensing properties of Ce-doped SnO_2 hollow spheres, *Sens. Actuators B* 173 (2012) 839– 846.

[10] J. Y. Patil, D. Y. Nadargi, J. Gurav, I. S. Mulla, S. Suryavanshi, Glycine combusted $ZnFe_2O_4$ gas sensor: Evaluation of structural, morphological and gas response properties, *Ceramics Intl.* 40 (2014) 10607–10613.

[11] Y. Lin, W. Wei, Y. Wang, J. Zhou, D. Sun, X. Zhang, S. Ruan, Highly stabilized and rapid sensing acetone sensor based on Au nanoparticle-decorated flower-like ZnO microstructures, *J. Alloys compd.* 650 (2015) 37-44.

[12] R. J. Wu, J. G. Wu, M. R. Yu, T. K. Tsai, C. T. Yeh, Promotive effect of CNT on Co_3O_4-SnO_2 in a semiconductor-type CO sensor working at room temperature, *Sens. Actuators B* 131 (2008) 306-312.

[13] B. Zhao, G. Shao, B. Fan, W. Li, X. Pian, R. Zhang, Enhanced electromagnetic wave absorption properties of NieSnO_2core shell composites synthesized by a simple hydrothermal method, *Mater. Lett.* 121 (2014) 118-121.

[14] M. Yin, S. Liu, Preparation of ZnO hollow spheres with different surface roughness and their enhanced gas sensing property, *Sens. Actuators B* 197 (2014) 58-65.

[15] N. Wu, M. Zhao, J. G. Zheng, C. Jiang, B. Myers, S. Li, M. Chyu, S. X. Mao, Porous CuO-ZnO nanocomposite for sensing electrode of high-temperature CO solid state electrochemical sensor, *Nanotechnol.*16 (2005), 2878-2881.

[16] S. K. Lim, S. H. Hwang, D. Chang, S. Kim, Preparation of mesoporous In_2O_3 nanofibers by electrospinning and their application as a CO gas sensor, *Sens. Actuators B* 149 (2010) 28-33.

[17] Hafeezullah, Z. H. Yamani, J. Iqbal, A. Qurashi, A. Hakeem, Rapid sonochemical synthesis of In_2O_3 nanoparticles their doping optical, electrical and hydrogen gas sensing properties, *J. Alloys Compd.* 616 (2014) 76-80.

[18] S. T. Navale, G. D. Khuspe, M. A. Chougule, V. B. Patil, Polypyrrole, α-Fe_2O_3 and their hybrid nanocomposite sensor: An impedance spectroscopy study, *Org. Electron.* 15 (2014) 2159-2167.

[19] B. Geng, F. Zhan, C. Fang, N. Yu, A facile coordination compound precursor route to controlled synthesis of Co_3O_4 nanostructures and their room temperature gas sensing properties, *J. Mater. Chem.* 18 (2008) 4977-4984.

[20] E. Sennik, N. Kilinc, Z. Z. Ozturk, Electrical and VOC sensing properties of anatase and rutile TiO_2 nanotubes, *J. Alloys Compd.* 616 (2014) 89-96.

[21] T. Xiao, X. Y. Wang, Z. H. Zhao, L. Li, L. Zhang, H. C. Yao, J. S. Wang, Z. J. Li, Highly sensitive and selective acetone sensor based on C-doped WO_3 for potential diagnosis of diabetes mellitus, *Sens. Actuators B* 199 (2014) 210-219.

[22] S. Wei, J. Zhao,W. Du, Synthesis, characterization and acetone-sensing properties of bristlegrass-like ZnO nanostructure, *Ceramics Intl.* 41(2015)769–776.

[23] W. Q. Li, S. Y. Ma, J. Luo, Y. Z. Mao, L. Cheng, D. J. Gengzang, X. L. Xu, S. H. Yan, Synthesis ofhollowSnO_2 nano-belts and their application in acetone sensor, *Mater. Lett.* 132 (2014) 338–341.

[24] S. B. Khan, M. Faisal, M. M. Rahman, A. Jamal, Low-temperature growth of ZnO nanoparticles: Photocatalyst and acetone sensor, *Talanta* 85 (2011) 943– 949.

[25] S. H. Yan, S. Y. Ma, X. L. Xu, W. Q. Li, J. Luo, W. X. Jin, T. T. Wang, X. H. Jiang, Y. Lu, H.S. Song, Preparation of SnO_2–ZnO hetero-nanofibers and their application in acetone sensing performance, *Mater. Lett.* 159 (2015) 447–450.

[26] M. M. Rahman, S. B. Khan, A. M. Asiri, K. A. Alamry, A. A. P. Khan, A. Khan, M. A. Rub, N. Azum, Acetone sensor based on solvothermally prepared ZnO doped with Co_3O_4 nanorods, *Microchim. Acta* 180 (2013) 675–685.

[27] M. M. Rahman, S. B. Khan, A. Jamal, M. Faisal, A. M. Asiri, Fabrication of highly sensitive acetone sensor based on sonochemically prepared as-grown Ag_2O nanostructures, *Chem. Engr. J.* 192 (2012) 122–128.

[28] M. M. Hussain, M. M. Rahman, A. M. Asiri. Ultrasensitive and selective 4-Aminophenol chemical sensor development based on nickel oxide nanoparticles decorated carbon nanotube nanocomposites for green environment. *J. Environ. Sci.* 53 (2017) 27-38.

[29] M. M. Hussain, M. M. Rahman, A. M. Asiri. M. R. Awual. Non-enzymatic simultaneous detection of L-glutamic acid and uric acid using mesoporous Co_3O_4 nanosheets. *RSC Adv.* 6 (2016) 80511-80521.

[30] M. M. Rahman, M. M. Hussain, A. M. Asiri. A novel approach towards the hydrazine sensor development by SrO. CNT nanocomposites. *RSC Adv.* 6 (2016) 65338-65348.

[31] R. Srivastava, Investigation on Temperature Sensing of Nanostructured Zinc Oxide Synthesized via Oxalate Route, *J. Sens. Technol.* 2 (2012) 8-12.

[32] M. M. Rahman, S. B. Khan, H. M. Marwani, A. M. Asiri, K. A. Alamry, M. A. Rub, A. Khan, A. A. P. Khan, N. Azuma, Facile synthesis of doped ZnO-CdO nanoblocks as solid-phase adsorbent

and efficient solar photo-catalyst applications, *J. Indust. Engr. Chem.* 20 (2014) 2278–2286.

[33] M. M. Hussain, M. M. Rahman, A. M. Asiri. Efficient 2-nitrophenol chemical sensor development based on Ce_2O_3 nanoparticles decorated CNT nanocomposites for environmental safety. *PLOS ONE* 11(2016) e0166265.

[34] M. M. Rahman, M. M. Hussain, A. M. Asiri. A glutathione biosensor based on a glassy carbon electrode modified with CdO nanoparticle-decorated carbon nanotube in a nafion matrix. *Microchim. Acta* 183 (2016) 3255-3263.

[35] M. M. Rahman, M. M. Hussain, A. M. Asiri. Fabrication of 3-methoxyphenol sensor based on Fe_3O_4 decorated carbon nanotube nanocomposites for environmental safety. *PLOS ONE* 12 (2017) e0177817.

[36] M. M. Rahman, M. M. Hussain, A. M. Asiri. Bilirubin sensor based on CuO-CdO composites deposited in a nafion/glassy carbon electrode matrixes. *Progress Nat. Sci: Mater. Intl.* 27 (2017) 566-573.

[37] Y. A. Hadeethia, A. Umarc, S. H. Al-Henitia, R. Kumare, S. H. Kim, X. Zhang, B.M. Raffah, 2D Sn-doped ZnO ultrathin nanosheet networks for enhanced acetone gas sensing application, *Ceramics Intl.* 43 (2017) 2418–2423.

[38] M. Sookhakian, Y. M. Amin, W. J. Basirun, Hierarchically ordered macro-mesoporous ZnS microsphere with reduced graphene oxide supporter for a highly efficient photo degradation of methylene blue, *Appl. Surface Sci.* 283 (2013) 668– 677.

[39] M. J. Akhtar, M. Ahamed, S. Kumar, M. A. M. Khan, J. Ahmad, S. A. Alrokayan, Zinc oxide nanoparticles selectively induce apoptosis in human cancer cells through reactive oxygen species, *Intl. J. Nanomed.* 7 (2012) 845–857.

[40] P. Fageria, S. Gangopadhyay, S. Pande, Synthesis of ZnO/Au and ZnO/Ag nanoparticles and their photocatalytic application using UV and visible light, *RSC Adv.* 4 (2014) 24962-24972.

[41] H. Wang, B. Li, J. Gao, M. Tang, H. Feng, J. Li, L. Guo, SnO_2 hollow nanospheres enclosed by single crystalline nanoparticles for highly efficient dye-sensitized solar cells, *Cryst. Eng. Comm.* 14 (2012) 5177–5181.

[42] X. T. Chen, K. X. Wang, Y. B. Zhai, H. J. Zhang, X. Y. Wu, X. Wei, J. S. Chen, A facile one-pot reduction method for the preparation of a SnO/SnO_2/GNS composite for high performance lithium ion batteries, *Dalton Trans.* 43 (2014) 3137-3143.

[43] M. V. Shestakov, V. K. Tikhomirov, D. Kirilenko, A. S. Kuznetsov, L. F. Chibotaru, A. N. Baranov, G. V. Tendeloo, V. V. Moshchalkov, Quantum cutting in Li (770 nm) and Yb (1000 nm) co-dopant emission bands by energy transfer from the ZnO nano-crystalline host, *Optics Express* 19 (2011) 15955-15964.

[44] R. Jenkins, R. L. Snyder, *Introduction to X-ray Powder Diffractometry*, John Wiley & Sons 138 (1994) 750-950.

[45] M. M. Rahman, M. M. Hussain, A. M. Asiri. Ultrasensitive and label-free detection of creatine based on CdO nanoparticles: A real sample approach. *New J. Chem.* 41 (2017) 6667-6677.

[46] M. M. Hussain, M. M. Rahman, A. M. Asiri. Sensitive L-Leucine sensor based on a glassy carbon electrode modifed with SrO nanorods. *Microchim. Acta* 183 (2016) 3265-3273.

[47] S. Sagadevan, J. Podder, Investigation on structural, surface morphological and dielectrical properties of Zn-doped ZnO_2 nanoparticles, *Mater. Res.* 19 (2016) 420-425.

[48] M. Arshad, M. M. Ansari, A. S. Ahmed, P. Tripathi, S. S. Z. Ashraf, A. H. Naqvi, A. Azam, Band gap engineering and enhanced photoluminescence of Mg doped ZnO nanoparticles synthesized by wet chemical route, *J. Luminescence* 161 (2015) 275–280.

[49] G. Sun, F. Qi, S. Zhang, Y. Li, Y. Wang, J. Cao, H. Bala, X. Wang, T. Jia, Z. Zhang, Synthesis and enhanced gas sensing properties of flower-like SnO_2 hierarchical structures decorated with discrete ZnO nanoparticles, *J. Alloys Compd.* 617 (2014) 192–199.

[50] S. Ameen, M. S. Akhtar, H. K. Seo, Y. S. Kim, H. S. Shin, Influence of Sn doping on ZnO nanostructures from nanoparticles to spindle

shape and their photo-electrochemical properties for dye sensitized solar cells, *Chem. Engr. J.* 187 (2012) 351– 356.

[51] D. Xu, D. Fan, W. Shen, Catalyst-free direct vapor-phase growth of $Zn_{1-x}Cu_xO$ micro-cross structures and their optical properties, *Nanoscale Res. Lett.* 8 (2013) 46.

[52] N. Gogurla, A. K. Sinha, S. Santra, S. Manna, S. K. Ray, Multifunctional Au-ZnO Plasmonic Nanostructures for Enhanced UV Photodetector and Room Temperature NO Sensing Devices, *Sci. Rep.* 4 (2014) 6483.

[53] S. S. Patil, M. G. Mali, M. S. Tamboli, D. R. Patil, M. V. Kulkarni, H. Yoon, H. Kim, S. S. Al-Deyab, S. S. Yoonc, S. S. Kolekarb, B. B. Kale, Green approach for hierarchical nanostructured Ag-ZnO and their photocatalytic performance under sunlight, *Catal. Today* 260 (2016) 126-134.

[54] Y. Qin, J. Xiong, W. Zhang, L. Liu, Y. Cui, H. Gu, Facile synthesis and photocatalytic performance of Mg_2SnO_4/SnO_2 heterostructures, *J. Mater. Sci.* 50 (2015) 5865–5872.

[55] X. T. Chen, K. X. Wang, Y. B. Zhai, H. J. Zhang, X. Y. Wu, X. Wei, J. S. Chen, A facile one-pot reduction method for the preparation of a SnO/SnO_2/GNS composite for high performance lithium ion batteries, *Dalton Trans.* 43 (2014) 3137-3143.

[56] K. He, Y. Luo, R. Yu, J. Qi, X. Sun, Y. Yang, H. Xu, J. Ma, D. Xu, Influence of Yb_2O_3 on electrical and microstructural characteristics of $CaCu_3Ti_4O_{12}$ ceramics, *Mater. Res. Bull.* 69 (2015) 98–103.

[57] Lee, S. J. Choi, K. M. Park, S. S. Lee, S. Choi, I. D. Kim, C. O. Park,The stability, sensitivity and response transients of ZnO, SnO_2 and WO_3 sensors under acetone, toluene and H_2S environments, *Sens. Actuators B* 197 (2014) 300–307.

[58] D. An, X. Tong, J. Liu, Q. Wang, Q. Zhou, J. Dong, Y. Li, Template-free hydrothermal synthesis of ZnO micro/nano-materials and their application in acetone sensing properties, *Superlattices Microstructures* 77 (2015) 1–11.

[59] P. P. Sahay, Zinc oxide thin film gas sensor for detection of acetone, *J. Mater. Sci.* 40 (2005) 4383–4385.

[60] M. Choudhury, S. S. Nath, D. Chakdar, G. Gope, R. K. Nath, Acetone sensing of ZnO quantum dots embedded in polyvinyl alcohol matrix, *Adv. Sci. Lett.* 3 (2010) 6–9.

[61] C. C. Wang, Y. C. Weng, T. C. Chou, Acetone sensor using lead foil as working electrode, *Sens. Actuators B* 122 (2007) 591–595.

[62] S. B. Khan, M. Faisal, M. M. Rahman, A. Jamal, Low-temperature growth of ZnO nanoparticles: Photocatalyst and acetone sensor, *Talanta* 85 (2011) 943– 949.

[63] M. M. Rahman, S. B. Khan, A. M. Asiri, K. A. Alamry, A. A. P. Khan, A. Khan, M. A. Rub, N. Azum, Acetone sensor based on solvothermally prepared ZnO doped with Co_3O_4 nanorods, *Microchim. Acta* 180 (2013) 675–685.

[64] M. M. Rahman, S. B. Khan, A. Jamal, M. Faisal, A. M. Asiri, Fabrication of highly sensitive acetone sensor based on sonochemically prepared as-grown Ag_2O nanostructures, *Chem. Engr. J.* 192 (2012) 122–128.

[65] C. C. Wang, Y. C. Weng, T. C. Chou, Acetone sensor using lead foil as working electrode, *Sens. Actuators B* 122 (2007) 591–595.

[66] E. Sotelo-Gonzalez, M. T. Fernandez-Argüelles, J. M. Costa-Fernandez, A. Sanz-Medel, Mn-doped ZnS quantum dots for the determination of acetone by phosphorescence attenuation, *Anal. Chem. Acta* 712 (2012) 120-126.

[67] M. M. Hussain, M. M. Rahman, M. N. Arshad, A. M. Asiri. Hg^{2+} sensor development based on (*E*)-*N′*-Nitrobenzylidine benzene sulfonohydrazide (NBBSH) derivatives fabricated on a glassy carbon electrode with a nafion matrix. *ACS Omega* 2 (2017) 420-431.

[68] M. M. Hussain, M. M. Rahman, M. N. Arshad, A. M. Asiri. Electrochemical detection of Ni^{2+} ions using synthesized (*E*)-*N′*-chlorobenzylidene-4-methylbenzenesulfonohydrazide derivatives modified with a nafion matrix. *Chemistry Select* 2 (2017) 7455-7464.

[69] M. M. Hussain, M. M. Rahman, M. N. Arshad, A. M. Asiri. Trivalent Y^{3+} ionic sensor development based on (*E*)-Methyl-*N′*-nitro-benzylidene-benzenesulphonohydrazide (MNBBSH) derivatives modified with nafion matrix. *Sci. Rep.* 7 (2017) 5832.

In: Acetone
Editors: Irenka Kozłowska et al.
ISBN: 978-1-53613-298-4

Chapter 6

A TECHNO-ECONOMIC ANALYSIS OF ISOPROPANOL-ACETONE-HYDROGEN CHEMICAL HEAT PUMP

***Pei Liu*[1,2,3]*, Jiwen Cen*[1,2]*, Yiwei Wang*[1,2] *and Fangming Jiang*[1,2,*]**

[1]Laboratory of Advanced Energy Systems, Guangzhou Institute of Energy Conversion, Chinese Academy of Sciences (CAS), Guangzhou, P.R. China

[2]CAS Key Laboratory of Renewable Energy, Guangzhou, P.R. China

[3]University of Chinese Academy of Sciences, Beijing, P.R. China

ABSTRACT

This paper presents an experimental system of isopropanol-acetone-hydrogen chemical heat pump (IAH CHP) for the recovery of low-grade (<100°C) industrial waste heat. Using coefficient of performance (COP) and exergy efficiency as evaluation indices, the system performance is systematically evaluated based on the experimental data. The

* Corresponding Author Email: jiangfm@ms.giec.ac.cn.

performance of the IAH CHP is found to be greatly affected by the endothermic and exothermic temperatures, and the flow rate of working fluid. In addition, the economic performance of the IAH CHP is analyzed as well. The results indicate that the long service-life of IAH CHP can significantly improve the system's economic performance, compensating for the high initial investment. The local energy price is found to have impact on the CHP's economic performance. If the local price of electricity is sufficiently high, the CHP may have better economic performance than other comparable energy systems.

Keywords: IAH CHP, heat pump, COP, exergy, economic performance

INTRODUCTION

With the continuous decline of the fossil fuels and to prevent the environment from deterioration, it appeals to develop alternative new energy systems to recover low-grade waste heat. Most of the waste heat released to the environment cannot be reutilized because of its low temperature, discontinuousness and the high cost for recovery. Heat pump, including mechanical heat pump, sorption heat pump and chemical heat pump, can transform low-grade heat that cannot be or is at least very difficult to be used into high-grade heat, so that part of the waste heat is recovered. Mechanical heat pump is widely used in daily life, but its shortcomings like high cost, limited temperature-upgrading range, and low efficiency make it hard to be applied for low-grade industrial waste heat recovery (Ito et al., 1991; Morten et al., 2012). Although sorption heat pump has a wide temperature-upgrading range, its large pressure shift leads to high operation cost (Hasan et al., 2008; Kitikiatsophon et al., 2004).

Chemical heat pump (CHP), in which chemical reaction plays a key role in transforming low-grade heat into high-grade heat and also in the conversion between chemical energy and heat, is proposed to overcome the limitations of mechanical and sorption heat pumps (Hasan et al., 2008). Some CHPs are introduced and classified in different ways (Ito et al., 1991). By adopting an appropriate reaction pair, the CHP may have a different energy density and temperature-upgrading range (Spoelstra et al., 2002; Kim et al., 1992; Prevost et al., 1980; KlinSoda et al., 2007; Gandia et al., 1992). In 1980, a CHP system using Isopropanol-Acetone-Hydrogen (IAH) as working pair was firstly proposed; since then, the possibility of using the dehydrogenation of isopropanol in a heat-driven CHP has been widely studied (Kitikiatsophon et al., 2004; Gandia et al., 1994; Ito et al., 1991; Taneda et al., 1994; Yamashita M., Saito Y.. 1993; Taneda et al., 1995; Cho et al., 2007; Cho et al., 2007).

As one of the most promising CHPs, the IAH CHP was studied mainly through simulation and experiment. A dynamic computer simulation of a micro-CHP system (Anikeev et al., 1993; Cunningham et al., 1996) demonstrated that the dynamics of micro-CHP can be successfully simulated and an economical analysis can be readily incorporated as a part of the dynamic simulation. Xu et al. (Xu et al., 2014) proposed an IAH-CHP system with two exothermic reactors in serial connection and found that the COP and exergy efficiency increased by 7.6% and 10.3%, respectively, relative to the conventional one that has only one high-temperature equilibrium reactor. Guo and Huai (Guo et al., 2012) designed a multi-parameter optimization recuperator in an IAH CHP system based on the entransy dissipation theory; they believed that the optimization design approach of heat exchanger based on entransy dissipation theory was very effective at both component and system levels.

The performance of the IAH CHP was found to be greatly affected by the endothermic and exothermic temperatures, and the flow rate of working fluid; the increase of reflux ratio, exothermic reaction temperature and endothermic reaction temperature leads to the decrease of enthalpy efficiency and exergy efficiency of IAH heat pump. (Guo et al., 2014; Xin et al., 2014) Theoretical analysis and experimental study of IAH CHP system are of paramount importance to the industrial use of IAH CHP. Nevertheless, previous studies on IAH CHP mainly focused on the dehydrogenation of isopropanol and hydrogenation of acetone by either simulation or experiment. In this paper, we conduct a comprehensive analysis of the system operating performance and economic cost. In order to get the optimal operation conditions, we build an experimental system with a newly designed reactive distillation column and an exothermic reactor. Based on experimental results, we evaluate and compare the economic performance of IAH CHP with some comparable energy systems. Some principles for the design and optimum operation of IAH CHP will be proposed as well.

IAH CHP SYSTEM

A schematic of the IAH CHP experimental system is shown in Figure 1. The test rig mainly consists of seven sections, including an endothermic reactor, a distillation column, a pre-heater, an exothermic reactor, a regenerator, a pump and data acquisition system.

The endothermic reactor is a stainless-steel tank being settled in a heating sleeve; an electric heater with temperature controller is used to mimic the waste heat. The mixture of isopropanol and Raney nickel catalyst was injected into the tank. Prevost and Bugarel (Prevost et al., 1980) proposed that the dehydrogenation of isopropanol in the endothermic reactor of IAH CHP existed mainly in the form of vapor, i.e.,

$$(CH_3)_2CHOH(l) \rightarrow (CH_3)_2CO(g) + H_2(g) \quad \Delta H_L = 100.4 kJ \cdot mol^{-1} \quad (1)$$

$$(CH_3)_2CHOH(l) \rightarrow (CH_3)_2CHOH(g) \quad \Delta H = 45.4 kJ \cdot mol^{-1} \quad (2)$$

where ΔH_L represents the endothermic reaction enthalpy.

The dehydrogenation of isopropanol takes place in the endothermic reactor and absorbs heat at temperature T_L to yield acetone and hydrogen. The products of the reaction, including acetone, hydrogen and unreacted isopropanol, are fed into the distillation column through the top flange.

The distillation column placed on the top of the reactor has a height of 1600 cm and a diameter of 10 cm. In the lower part of the column a port is settled, through which some Raney nickel catalyst is fed into the distillation column to further the dehydrogenation reaction of isopropanol.

The isopropanol is separated from the mixture fed back to the endothermic reactor. The acetone and hydrogen as the product are driven by the pump, and then flow to the regenerator. Subsequently, the acetone and hydrogen are introduced into the exothermic reactor after being heated up to the reaction temperature in the pre-heater. The pre-heater is an electric heater, and it only takes effect when the system starts.

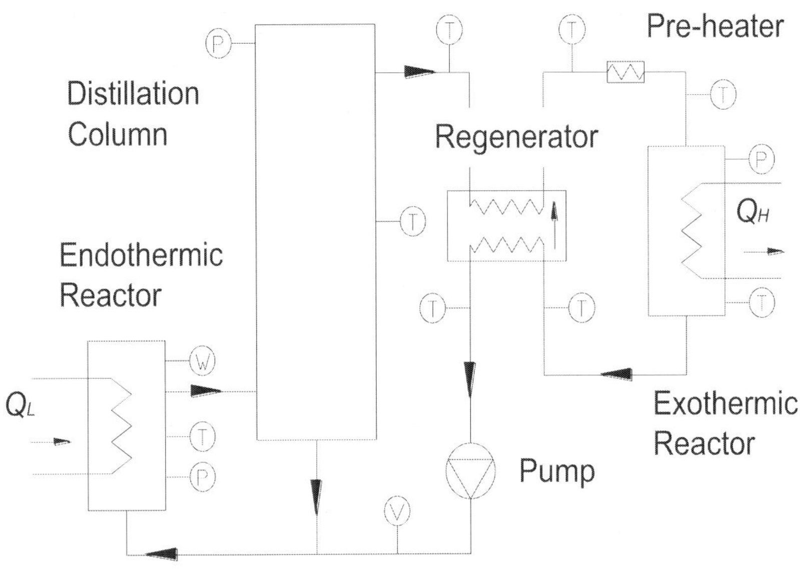

Figure 1. Schematic of the IAH CHP and the data acquisition system.
P: pressure sensor, T: thermocouple, V: flow meter, W: voltage & current meter.

The exothermic reactor is also a stainless-steel tank with some Raney nickel catalyst fed inside. The hydrogenation of acetone takes place in the exothermic reactor in the gas phase at temperature T_H. The acetone and hydrogen fed into the reactor combine to produce isopropanol and release heat at temperature T_H. The product isopropanol is then introduced to the regenerator to release heat and subsequently pumped back to the endothermic reactor to close a cycle. The hydrogenation reaction of acetone takes place in the exothermic reactor can be expressed as following:

$$(CH_3)_2CO(g)+H_2(g)\rightarrow(CH_3)_2CHOH(g)\ \ \Delta H_H=-55.0kJ\cdot mol^{-1} \quad (3)$$

The regenerator is a shell-and-tube heat exchanger that is also made of stainless-steel. The mixture from distillation column is heated up by the product from exothermic reactor to a higher temperature close to the point that can trigger the hydrogenation reaction of acetone. The heated mixture goes through the de-functioned pre-heater and enters the exothermic reactor. All the components and line of the system are insulated to reduce the heat loss.

Temperature including endothermic and exothermic temperature is the key parameter affecting the system performance. In the experiment, some operation cases are set to study the characteristics of the IAH CHP cycle: the heater is designed for 3 kW systems; endothermic temperature is set from 343K to 393K by adjusting the tank temperature. In each test case, data is recorded after stable running for 15 minutes. Then we can get the COP and exergy data to evaluate the performance.

FIDELITY ANALYSIS

The status parameters of some key components are measured to evaluate the performance of the IAH. Temperature can be used to calculate the amount of heat transmitted by the system. Flow rate and pressure can be used to study the optimum running state of the system. Therefore, the IAH CHP system was instrumented to measure the parameters of interest; locations of monitoring points are shown in Figure 1. Temperature is measured with K-type thermocouples with accuracy of ±0.5 K. There are 9 temperature monitoring points in total, positioned in the endothermic and exothermic reactor, at the inlet and outlet of the regenerator and the pre-heater, on the surface of the column, endothermic and exothermic reactor, and the environment, respectively. The power consumed by the electric heater (including the pre-heater) and pump is measured with the electrical voltage and current meter, and the measurement accuracy is 0.5%. The pressure in the endothermic reactor, distillation column and the exothermic reactor is measured with pressure sensors which have accuracy of 0.5%. The flow rate is measured with a flow meter which has 4.0% accuracy. And two sampling holes are set for chromatographic analysis at the inlet and outlet of the exothermic reactor.

According to the Coleman and Steele method (Coleman et al., 1989) and the ANSI/ASME standard (ANSI/ASME, 1986), if a parameter *R* is a function of some measured parameters (*Xi*), then the uncertainty of *R* is affected by the individual uncertainties of *Xi*, and the absolute and relative uncertainties of *R* can be got, respectively. The measurement error of the COP and exergy can be obtained; the relative uncertainty of COP and exergy are calculated as 8.36% and 6.52%, respectively.

The fidelity of the experiment is analyzed to ensure the reliability of the experiment. Taking the endothermic reactor as an example, the energy flow is shown as Figure 2. The energy input into the component is supplied by the electric heater. The energy output includes the heat loss through the surface and the chemical energy from the dehydrogenation. The energy input and output can be got from:

$$Q_{in} = U \cdot I \tag{4}$$

$$Q_{out} = Q_C + h \cdot A \cdot \Delta T \tag{5}$$

$$Q_C = v \cdot w \cdot \Delta H_L \tag{6}$$

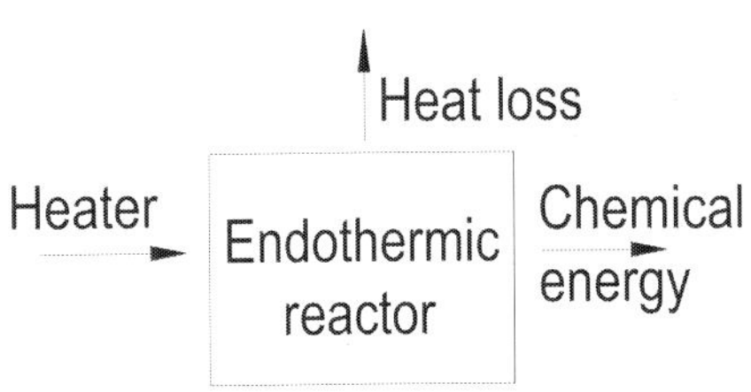

Figure 2. Energy flow of component (exemplified with endothermic reactor).

Table 1. System components parameters (exemplified with the endothermic temperature 85℃ & flow rate 1.47ml/s)

Component	Energy Flow	Fidelity %	Efficiency %
Endothermic reactor 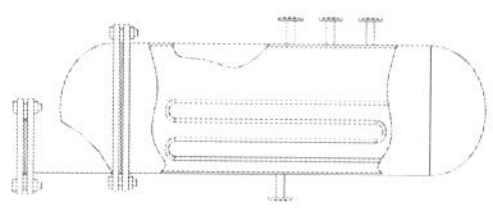	Input: $Q_{in} = U \cdot I$ Output: $Q_{out} = Q_C + h \cdot A \cdot \Delta T$	90.75	82.9
Regenerator 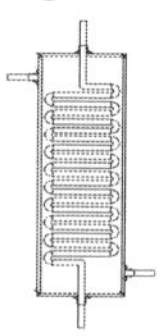	Input: $Q_{in} = Q_C$ Output: $Q_{out} = Q_C' + h \cdot A \cdot \Delta T$	94.55	78.6
Exothermic reactor 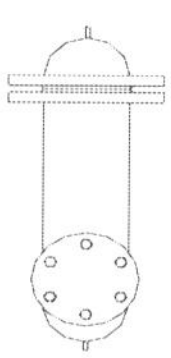	Input: $Q_{in} = Q_C$ Output: $Q_{out} = Q_C' + h \cdot A \cdot \Delta T$		93.3
Pump 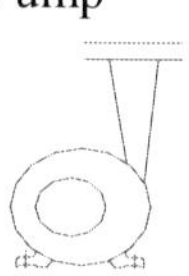	Input: $Q_{in} = U \cdot I + Q_C$ Output: $Q_{out} = E_H + h \cdot A \cdot \Delta T$		98.2

Here, *U* and *I* represent the voltage and current respectively, *h* represents the heat transfer coefficient, *A* represents the heat transfer area, *ΔT* represents the temperature difference, Q_C represents the chemical energy that absorbed in the reaction, *v* and *w* represent the flow rate and content respectively.

The energy must be conservative, with the meteorological chromatographic analysis and the Eq.1 and 2; we can get the system fidelity. The calculated flow rate is 1.62 ml/s, while the measured flow rate is 1.47 ml/s; as a result the fidelity is 90.75%. So we think that the result of the experiment is reliable. Similarly, the other components' system fidelity can be obtained and are shown in Table 1.

Here, Q_C and Q_C' represent the chemical energy that absorbed and released in the reaction and E_H represents the potential energy.

Also we get the efficiency of the component.

$$\delta = \frac{M_{act}}{M_{th}} \tag{7}$$

$$e = 1 - \frac{Q_{out}}{Q_{in}} \tag{8}$$

where δ represents the fidelity, M_{act} represents actual values, M_{th} represents the theoretical values, and e represents the efficiency.

THERMODYNAMIC ANALYSIS AND THERMAL-ECONOMIC ANALYSIS

The COP can be expressed as:

$$COP = \frac{Q_H}{Q_L + W_t} \tag{9}$$

$$Q_L = U \cdot I - h \cdot A \cdot \Delta T \tag{10}$$

$$Q_H = v \cdot \alpha_H \cdot \Delta H_H \tag{11}$$

where W_t represents the energy consumed by the pump, Q_L represents the waste heat, Q_H represents the heating capacity, α_H represents the reaction rate of the hydrogenation.

Exergy (E_X) is a concept in the thermodynamics field of physics that has come to replace enthalpy as a more accurate representation when it is used in an open system. Exergy represents the amount of energy present in a system that is available for useful work. Exergy efficiency (η) is a parameter that is used to evaluate the efficiency of a process by taking the second law of thermodynamics into account.

The exergy is defined as:

$$E_{X,Q} = \left(1 - \frac{T_0}{T}\right) \cdot \delta Q \tag{12}$$

$$\eta = \frac{E_{X,out}}{E_{x,in}} = \frac{\left(1 - \frac{T_0}{T_H}\right) \cdot \delta Q_H}{\left(1 - \frac{T_0}{T_L}\right) \cdot \delta Q_L + W} \tag{13}$$

where T_0 represents the environmental temperature, E_X and η represent the heating exergy and exergy efficiency provided by the system. A thermal-economic model, as schematically displayed in Figure 3 (Zhu et al., 1988), was chosen to perform an economic analysis of the IAH CHP. The model is established based on the analysis of exergy/heat produced and consumed. Figure 3 exemplifies the flow diagram of exergy.

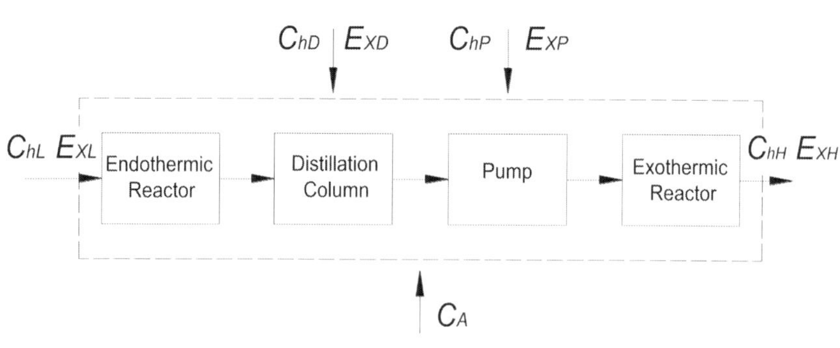

Figure 3. Thermo economics model of IAH-CHP (exemplified with the flow diagram of exergy).

Applying Eqs. (14) - (16), the exergy cost can be obtained, and then the heating cost can be obtained by replacing E_X with Q. The thermal-economic equation is:

$$C_{hH}E_{X\,hH} = C_{hL}E_{X\,hL} + C_{hD}E_{X\,hD} + C_{P}E_{X\,hP} + C_{A} \tag{14}$$

with,

$$C_{A} = \left\{\left[C_{0} - S_{v} \times PWF(i,n)\right] + \sum F_{m} \times PWF(i,n)\right\} \times CRF(i,n) \tag{15}$$

The exergy and heat costs of a thermal system used for heating and cooling are:

$$C_{hH} = \left(C_{A} + C_{hL}E_{X\,hL} + C_{hD}E_{X\,hD} + C_{hP}E_{X\,hP}\right) / E_{X\,hH} \tag{16}$$

where:

- C_A: system cost on a yearly base
- C_0: the initial investment
- S_v: system residual value
- F_m: operating cost in Year m
- $PWF(i,n)$: cash factor
- $CRF(i,n)$: capital recovery factor
- i: interest rate
- n: years
- T: ambient temperature, 293.15K
- E_X: exergy
- E_{XL}: heat (exergy) input to the endothermic reactor
- E_{XD}: heat (exergy) input to the distillation column
- E_{XP}: heat (exergy) input to the pump
- E_{XH}: heat (exergy) released from the exothermic reactor
- C_h: heating cost

RESULTS AND DISCUSSION

The endothermic temperature T_L, the exothermic temperature T_H, and fluid flow all influence the heat production, COP and exergy efficiency significantly. At a given running condition, we study the heat production, COP and exergy efficiency as a function of T_L, T_H and v.

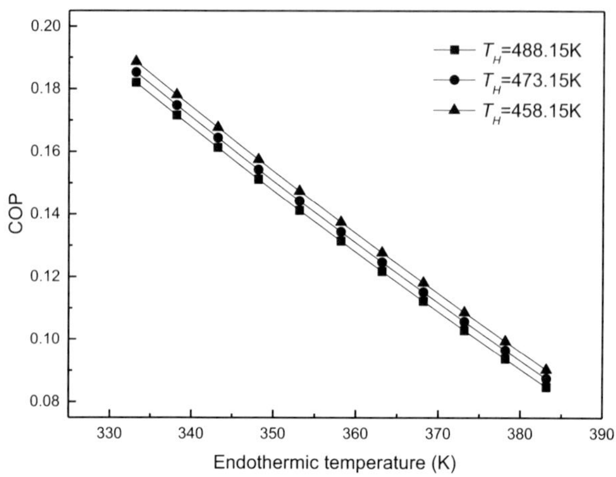

Figure 4. COP with various endothermic temperatures.

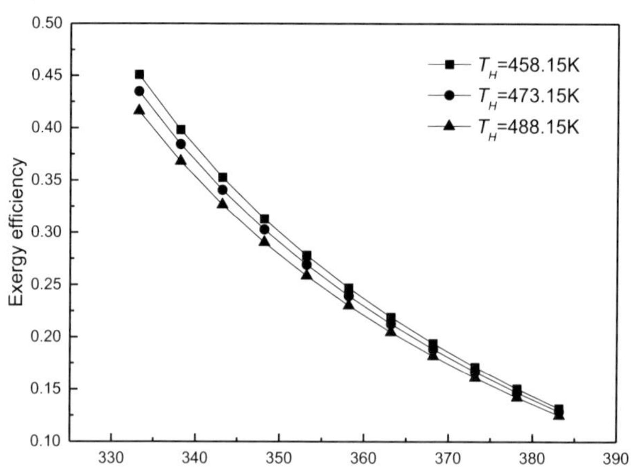

Figure 5. Exergy efficiency with various endothermic temperatures.

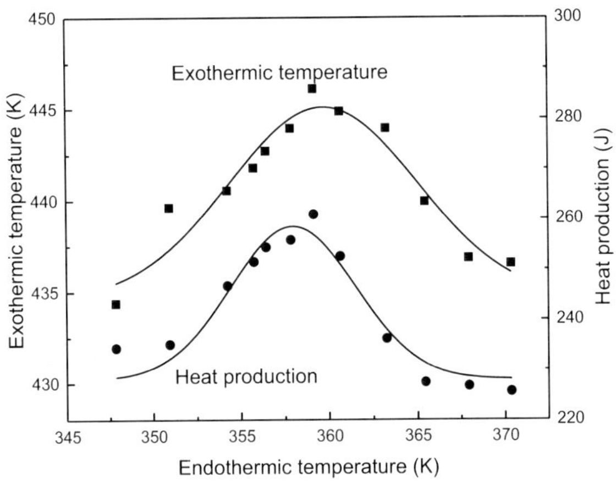

Figure 6. Exothermic temperature and heat production with various endothermic temperatures.

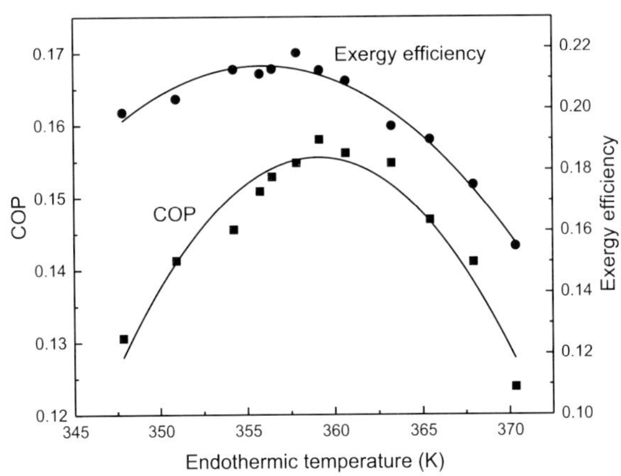

Figure 7. COP and exergy efficiency with various endothermic temperatures.

Firstly, the relationship between the temperature and COP, also between temperature and exergy efficiency are calculated, as shown in Figures 4 and 5. And then, in order to compare with the theoretical analysis results, some experiments are carried out on the IAH heat pump experimental system which is established on the basis of previous analysis. At a given endothermic temperature or fluid flow, the influences of exothermic temperature and flow on heat production, COP and exergy efficiency, are studied respectively and the variations are shown in Figures 6-9.

As shown in Figure 4, COP decreases as the endothermic reaction and exothermic reaction temperature increase. The cause is not only the effect of temperature on reaction rate, but also the changes of the pressure. The higher the endothermic temperature, the larger the isopropanol vapor enthalpy is and the higher the vapor pressure is. The vapor in the endothermic reactor consumes more energy from the heater as the pressure increases. And also higher endothermic temperature drives more mixture into the distillation column including the unreacted isopropanol. The increased isopropanol in the production improves the content, and absorbs the heat but with no dehydrogenation to convert the energy. All these lead to larger Q_L; therefore, the COP decreases. But in the subsequent experimental analysis, the variations are different from which are shown in Figures 6 and 7. As the endothermic temperature increases, exothermic temperature, heat production, COP and exergy efficiency all increase and reach the maximum at the endothermic temperature of about 360K, then decrease as the endothermic temperature further increases. The main reason for this variation results from the distillation column in the experimental system. There is a decrease in temperature due to the heat loss from the distillation column, which is not considered in numerical calculations. Endothermic reactions take place mainly at 353K, but the temperature in the distillation column is lower than that in the endothermic reactor. It needs more heat to keep the reaction in the distillation column. So the exothermic temperature, heat production and COP increases before the temperature rises to 360K. When T_L is above 360K, the temperature in the distillation column is above the boiling point of isopropanol. The vapor pressure and isopropanol content in the production from the distillation column increases with the increase of T_L. The increased vapor pressure and unreacted isopropanol in the production consume more energy from the heater. Both lead to more energy loss and larger Q_L. And the increased unreacted isopropanol driven to the exothermic reactor consumes more energy from the exothermic reaction without supplying heat to the outside.

Therefore, the higher the temperature, the lower the COP, the exothermic temperature and the heat production is.

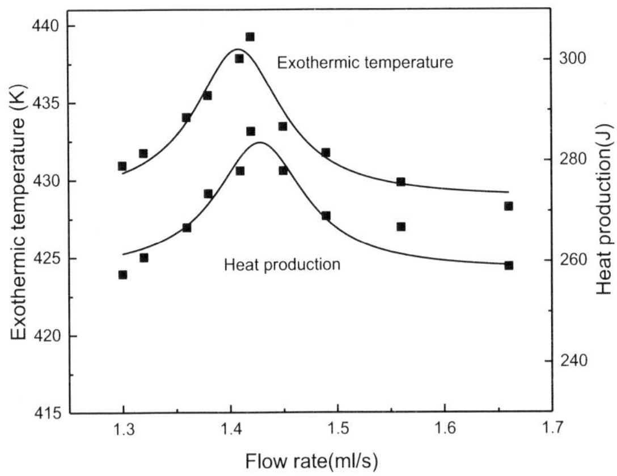

Figure 8. Exothermic temperature and heat production with various flow rates.

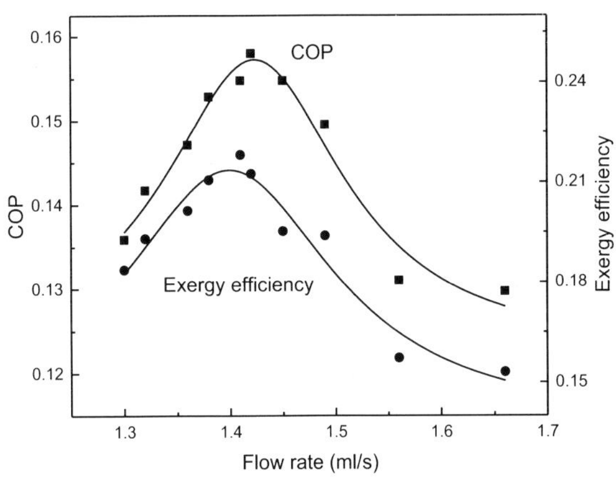

Figure 9. COP and exergy efficiency with various flow rates.

The fluid flow from the distillation column influences the heat production, COP and exergy efficiency significantly. Figures 8 and 9 display the influence of flow at a given endothermic temperature. It has a very similar pattern to the influence of endothermic temperature as discussed above. The exothermic temperature, heat production, COP and exergy efficiency all reach the maximum point at a flow rate of 1.4 ml/s. The fluid flow increases by adjusting heating power and the internal pressure of the distillation column. Under the condition of keeping the purity of the production from the distillation column, the endothermic temperature, exothermic temperature, heat production, COP and exergy efficiency would increase with the flow. In the actual experiment process, the purity would decrease after the flow reaches about 1.40 ml/s. With the decrease of the purity, more unreacted isopropanol in the production consumes more energy from the heater and the exothermic reaction without supplying heat to the outside. And then the exothermic temperature, heat production, COP and exergy efficiency would decrease with the flow.A comparison of IAH CHP with some other heat pumps including traditional mechanical heat pump, transcritical CO_2 heat pump, and coal-fired boiler are presented in Table 2. The IAH CHP has a 6.4 years payback period which is much longer than that of the mechanical heat pump and coal-fired boiler. But the mechanical heat pump can only provide heat below 373K, which can not be used directly in related industrial processes. Therefore, the heating costs are different for these different systems.

Table 2. System parameters

Type	IAH CHP	Mechanical HP	CO_2 HP	Coal-fired boiler
Heating power (kW)	10			
Source temperature (K)	353	288~298	273~298	none
Heating temperature (K)	473.15	328.15	353.15	673.15
Initial investment (yuan)	400000[a]	20000[a]	50000[a]	100000[a]
Life expectancy (year)	30	15	15	30
Energy consumption	Waste heat	Electricity		Coal
Energy price	0	0.8 yuan/kWh(ASKCI Consulting, 2017)		700 yuan/t(China Coal Industry Association, 2017)
Payback period (year)	6.4	0.48	1.20	2.3

[a] Provided by the design institute.

The exergy and heat costs of the IAH CHP, the mechanical heat pump, the transcritical CO_2 heat pump and the coal-fired boiler are shown in Figures 10 and 11. The electricity price is considered to be 0.8 yuan/kWh, and the system lifetime is considered to be 15 years for mechanical heat pumps and IAH CHP, and 30 years for coal-fired boiler, respectively. Since the temperature-lift is much larger, the exergy cost of IAH CHP is much lower than that of the mechanical heat pump.

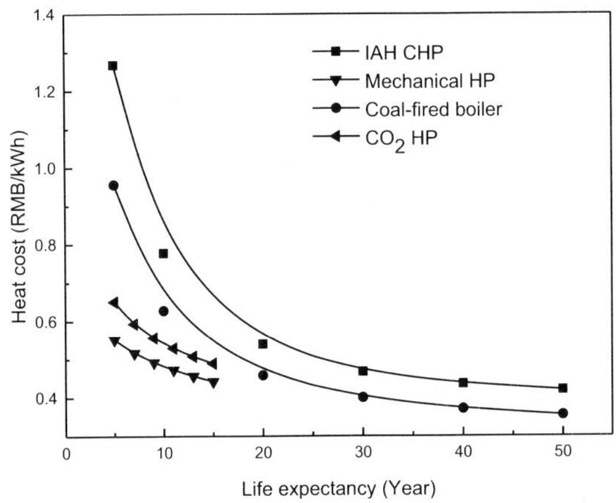

Figure 10. Heat cost with various life expectancy year.

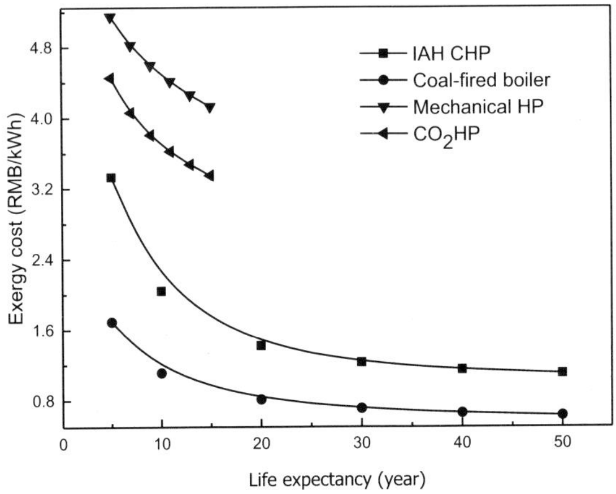

Figure 11. Exergy cost with various life expectancy year.

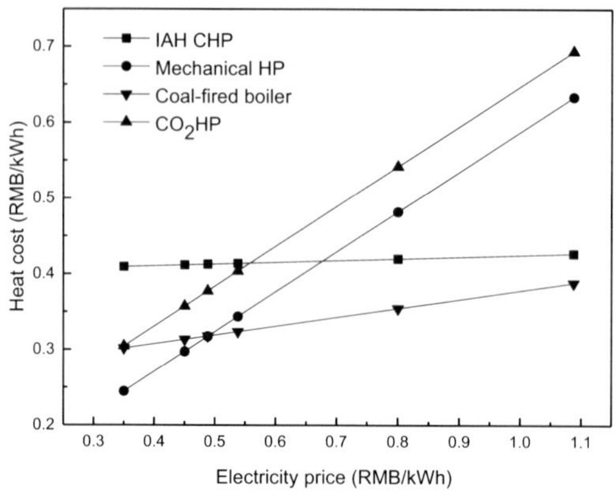

Figure 12. Heat cost with various electricity price.

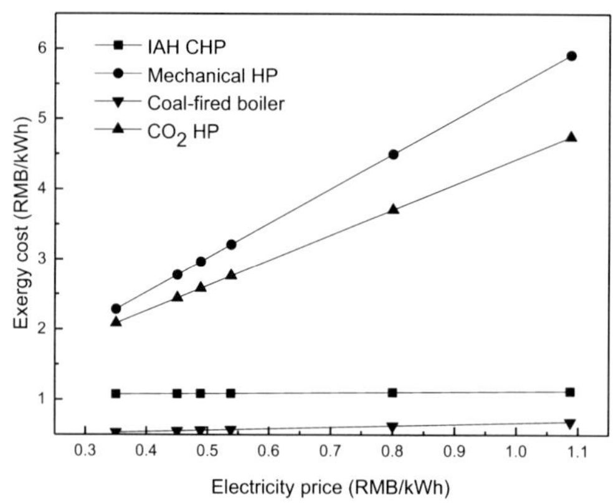

Figure 13. Exergy cost with various electricity price.

The effects of the expected lifetime and electricity price on the exergy and heat costs of different systems were studied also. Figures 11 and 12 display the calculated exergy and heat cost, respectively, as a function of the expected lifetime. The electricity price is considered to be 0.8 RMB/kWh in the calculations. Both the exergy and heat costs decrease with the increase of the system lifetime. The IAH CHP initial investment is much higher due to the expensive distillation column, but its annual operating cost is much lower as it needs little external energy consumption.

Therefore, the exergy and heat costs of IAH CHP are relatively high if the expected lifetime is short. However, with a longer expected lifetime (>10 years), the IAH CHP can have comparable exergy and heat costs to the other thermal systems. Seen from Figures 10 and 11, the exergy and heat cost of IAH CHP becomes even lower than that of the mechanical heat pump and the transcritical CO_2 heat pump when its lifetime is longer than 8 years or so.

Figures 12 and 13 present the calculated exergy and heat costs as a function of the electricity price for the four thermal systems. The lifetime of the thermal systems was considered to be 15 years for mechanical heat pumps and 30 years for IAH CHP and coal-fired boiler and the yearly working time was considered to be 7,800 hours in the calculations. IAH CHP requires almost no fuel or electricity consumption, and its exergy and heat costs show little dependence on the electricity price. Seen from Figures 12 and 13, the exergy and heat costs of the thermal systems, except for IAH CHP, increase with the increase of the electricity price. Since the mechanical heat pump directly consumes electricity to drive the compressor, its exergy and heat costs show much stronger dependence on the electricity price. From these two figures, it could be thought that IAH CHP becomes more advantageous if the electricity price is sufficiently high. Moreover, IAH CHP can provide cold when providing heat, further protruding its advantage.

CONCLUSION

This work was undertaken to investigate the techno-economic feasibility of Isopropanol-Acetone-Hydrogen CHP. The influence of endothermic and exothermic temperatures, isopropanol content and flow on the system performance has been studied by calculation and experiment. The comparison of heat and exergy cost between IAH CHP and some other traditional heating systems was carried out, respectively. The following conclusions can be drawn from this study:

1. The endothermic temperature in the reactor and distillation column has significant effect on hydrogen production and on the separation of acetone and isopropanol. The product flow rate increases with the increasing temperature, which means a higher system performance. The system performance reaches the maximum at about 360 K and 1.40 ml/s in the actual experimental system.
2. The design and heat insulation of the distillation column is important to the performance: a better design is beneficial to the fluid separation and higher reaction rate. Good heat insulation can reduce the heat loss and energy consumption.
3. Compared with other traditional heating systems, IAH CHP is promising: it has a greater advantage of lower heating cost with a longer life expectancy and higher local energy price.

ACKNOWLEDGMENTS

The research is supported by the National Science Foundation of China under grant No. (2011CB710700).

REFERENCES

ANSI/ASME, 1986. *Measurement Uncertainty*, PTC 19.1-1985.

ASKCI Consulting, 2017. *Business data.*

Anikeev V. I., Ooudkov A. V., Bobrin A. S., (1993). Study of catalytic heat pumps of upgrading low-level thermal energy, in: *New Energy Systems and Conversions*, Universal Academy Press Inc.:685-689.

China Coal Industry Association, 2017.

Cho H., Luck R., Chamra L.M., (20070. Dynamic Simulation of a Micro-CHP Facility: A Case Study, ASME Paper No. ES2007-36156,

Proceedings of 2007 Energy Sustainability Conference, Long Beach, CA, USA.

Cho H., Luck R., Chamra L.M., (2007). Technical and Economical Analysis of a Micro-CHP Facility Based on Dynamic Simulation: A Case Study, ASME Paper No. IMECE2007-41885, *Proceedings of 2007 ASME International Mechanical Engineering Congress and Exposition*, Seattle, WA, USA.

Cunningham J., Hickey J.N., Wang Z., (1996). Activities and selectivities of copper/metal-oxide catalysts at temperatures relevant to chemical heat pumps based on isopropanol/acetone interconversions, *Int. J. Energ. Re.* 20, 763-766.

Coleman H.W., Steele W.G., (1989). *Experimental and Uncertainty Analysis for Engineers*, Wiley, New York.

Guo J.F., Huai X.L., (2012). Optimization design of recuperator in a chemical heat pump system based on entransy dissipation theory, *Energy*, 41, 335-343.

Guo J.F., Huai X.L., Xu M., (2014). Study on Isopropanol-Acetone-Hydrogen chemical heat pump of storage type, *Sol. Energy*. 110, 684-690.

Gandia L.M., Montes M. (1992). Effect of the design variables on the energy performance and size parameters of a heat solar energy transformer based on the system acetone/H2/2-propanol, *Int. J. Energ. Re.* 16, 851-864.

Gandia L.M., Montes M., (1994). Effect of the reduction of temperature on the selectivity of the high temperature reaction of acetone and hydrogen over alumina and titania supported nickel and cobalt catalysts, *J. Mol. Catal.* 94, 347-367.

Hasan D., Moghtada M., Semra Ü., (2008). A review on adsorption heat pump: problems and solutions. *Renew. Sust. Energ. Rev.* 12(9),2381-403.

Ito E., Yamashita M., Hagiwara S., (1991). A composite Ru-Pt catalyst for 2-propanol dehydrogenation adaptable to the chemical heat pump system, *Chemistry Letters from the Chemical Society of Japan*, 351-354.

KlinSoda I., Piumsomboon P., (2007). Isopropanol-acetone-hydrogen chemical heat pump: a demonstration unit. *Energ.Convers.Manage.* 48,1200-1207.

Kitikiatsophon W., Piumsomboon P., (2004). Dynamic simulation and control of an Isopropanol-Acetone-Hydrogen chemical heat pump. *Science Asia*, 30, 135-47.

Kim T.G., Yeo Y.K., Song H.K., (1992). Chemical heat pump based on dehydrogenation and hydrogenation of i-propanol and acetone, *Int. J. Energ. Re. 16*, 897-916.

Morten B. Blarke, (2012). Towards an intermittency-friendly energy system: comparing electric boilers and heat pumps in distributed cogeneration. *Appl. Energ.* 91(1),349-65.

Prevost M., Bugarel R., (1980). Chemical heat pump: system isopropanol-acetone-hydrogen, in: *Proceedings of the International Seminar on Thermochemical Energy Storage*, Stockholm, 95-111.

Spoelstra S., Haije W.G., Dijkstra J.W., (2002). Techno-economic feasibility of high temperature high-lift chemical heat pumps for upgrading industrial waste heat, *Appl. Therm. Eng.* 22, 1619-1630.

Taneda, Daisuke, Yasutomi, Isamu, Shibata,Setsuo, (1994). Studies of 2-propanol/acetone/hydrogen energy conversion system, in: *10th World Hydrogen Energy Conference*, Cocoa Beach, FL, USA.

Wongsuwan W., Kumar S., Neveu P., Meunier F., (2001). A review of chemical heat pump technology and applications, *Appl. Therm. Eng.* 21, 1489-1519.

Xu M., Duan Y.J., Xin F., Huai X.L., (2014). Design of an isopropanol-acetone-hydrogen chemical heat pump with exothermic reactors in series. *Appl. Therm. Eng.* 71, 445-449.

Xin F., Xu M., Huai X.L., Li X.F., (2014). Characteristic and kinetic of liquid-phase isopropanol dehydrogenation over raney nickel catalysts for chemical heat pump, *Appl. Therm. Eng.* 70, 580-585.

Yamashita M., Saito Y., (1993). Catalyst study on liquid phase dehydrogenation of 2-propanol for newly-proposed chemical heat pump, in: *International Symposium on CO_2-fixation and efficient utilization of energy*, Tokyo, Japan.

Taneda, Daisuke, Yasutomi, (1995). Pilot plant proving tests of solar chemical heat pump system, in: *Solar Engineering*, ASME, 87-92.

Zhu M.S., Wang B.X., (1988). *Exergy analysis of energy systems*, Peking, Tsinghua University Press.

In: Acetone
Editors: Irenka Kozłowska et al.
ISBN: 978-1-53613-298-4

Chapter 7

Acetone as a Recyclable Solvent

Dale G. Drueckhammer*
Department of Chemistry, Stony Brook University,
Stony Brook, NY, US

Abstract

Acetone has a longstanding use in organic chemistry labs as the solvent of choice for the cleaning of glassware. This application extends to biological labs, where acetone may be used to rinse away organic contaminants, or water following water-based cleaning. Acetone waste from glassware cleaning can be readily recycled by fractional distillation. Its low boiling point (56°C) permits its effective separation from water and other high boiling substances, while small amounts of other volatile organics do not usually interfere with its repeated use in cleaning. Acetone has generally not been the solvent of choice for other applications in organic chemistry including chromatography on silica gel, where ethyl acetate in combination with hexane or other saturated hydrocarbon(s) is typically used as the mobile phase. Despite its limited use, acetone has certain advantages over the commonly used ethyl acetate in chromatography. First, it is typically less expensive. While acetone is generally considered to be UV transparent only above about 300 nm, it

* Corresponding Author Email: dale.drueckhammer@stonybrook.edu.

actually has a window of low absorbance in the 205 to 220 nm range. In contrast, ethyl acetate absorbs strongly up to about 250 nm. Acetone as solvent along with a non-absorbing saturated hydrocarbon co-solvent permits UV detection of eluent from chromatography at or near 210 nm, a much shorter wavelength that possible with ethyl acetate. This allows UV detection of a wider range of eluents, including simple alkenes. Acetone is generally considered to have a slightly greater eluting power than ethyl acetate in chromatography on silica gel. Acetone is an especially attractive alternative to ethyl acetate for chromatography if one wishes to recover and separate the solvent mixture by distillation, due to its lower boiling point (56°C) vs 77°C for ethyl acetate. Use of acetone in conjunction with heptane is especially practical, as the two solvents can be separated sufficiently for repeated use by fractional distillation using a standard automated commercial distillation instrument. This application of acetone has significant potential for reducing solvent consumption and waste generation in organic chemistry labs.

Keywords: solvent recycling, chromatography, fractional distillation

INTRODUCTION

In most organic chemistry lab environments, the largest quantity of chemical used and waste generated can be attributed to solvents. Between 80 and 90% of the total mass used in production of active pharmaceutical ingredients has been attributed to solvent [1] The annual waste generated in pharmaceutical industry has been estimated at 88 million kg, with 83% of this waste attributed to the top 10 solvents in use in pharmaceutical industry.[1] Information about amounts of solvents used and waste generated in academic labs is less readily available. However solvent undoubtedly dominates the waste stream of academic synthetic labs as well.

The largest component of the waste stream from medicinal chemistry is said to be solvent from chromatographic purification [2]. The technique of "flash" chromatography in which pressure is used to increase the flow rate through the column was described by Still and coworkers in a classic 1978 paper [3]. The widespread use of this technique is illustrated by the more than 8,000 citations of this article, including almost 60 in the first 10 months of 2017. This paper prescribes use of a volume of eluent of about three to four times the volume of the silica bed, with larger volume ratios used with the smallest columns. It is stated that increased resolution can be achieved using a less polar solvent mixture, in which case a larger volume of solvent is required for elution. As described in the Still reference, mixtures of ethyl acetate with hexane or other saturated hydrocarbon are most commonly used as solvent for flash chromatography, with the desired polarity achieved by the appropriate ratio of the two.

A detailed analysis of the costs associated with preparative chromatography showed that the largest contributor to cost was solvent vs. labor, equipment, packing, etc., and that solvent often contributes more than half of the total cost [4]. An earlier analysis attributed more than 80% of the cost to solvent [5]. In addition to solvent purchase cost, waste disposal costs and the environmental impact of large scale production and disposal of solvents has also been recognized. Several of the principles of green chemistry are especially relevant to preparative chromatography, including the prevention or minimization of waste as well as impact on safety, health, and environment [6]. Sustainable chemistry would appear to require a reduction in solvent use and the use of "greener" solvents [7]. While both economic and environmental concerns would appear to require recycling and reuse of solvents in large scale industrial processes, solvent recycling has not been viewed as feasible in a medicinal chemistry or discovery research environment [8].

In addition to the use of ethyl acetate-hexane mixtures for purifications by chromatography on silica gel, industrial medicinal chemistry labs and organic chemistry labs in general often use dichloromethane-methanol mixtures in purification of more polar compounds [2]. Dichloromethane is toxic and persists in the environment when released. Dichloromethane is also nonflammable and thus cannot be disposed of by incineration, as is done with other organic solvents [2]. Several solvent selections guides have been developed based on environmental friendliness/compatibility [9]. One focus of these efforts has been to replace dichloromethane and other halogenated solvents with nonhalogenated alternatives [2]. These evaluations have generally been performed with the expectation that used solvent will be disposed of, with attention usually not given to possibilities of solvent recycling.

In addition to efforts to decrease the environmental impact of solvent use by changing over to more environmentally friendly solvents, efforts have also been made to reduce the volume of solvent waste generated by chromatography. This includes strategies to simply reduce the volume of solvent employed in a given chromatographic separation as well as seeking alternatives to silica-based chromatography and methods to avoid chromatography entirely [8]. The volume of solvent can be reduced by using the smallest column that is sufficient, or using a smaller column packed with higher quality silica gel. TLC studies to identify the proper solvent for elution can reduce solvent use relative to running a long gradient from low to high polarity. Stopping elution as soon as the desired product(s) are eluted can also reduce excess solvent use. Alternatives to chromatography on silica gel include super-critical fluid chromatography [6], liquid-liquid chromatography with ionic liquids [10], and ion exchange and reverse phase chromatography [8]. Carrying sufficiently pure products to the next step without chromatography, and alternative purification methods such as crystallization have been suggested to reduce solvent use [8]. Such strategies are more likely to be feasible in development of process/production applications than in discovery chemistry where speed and time/labor efficiency may take priority over environmental and sustainability considerations.

In recent years, commercial automated chromatography systems have come into common use. In addition to an automated programmable pumping system, these instruments include automatic fraction collection and generally UV detection, though other types of detection are also available. The default parameters for these systems usually provide for even larger solvent volumes to be used, with the expectation of using a less polar solvent mixture to achieve greater resolution. Indeed the automation of these systems greatly increases the convenience of larger volume runs in contrast to manual setups where the user is required to continually apply pressure, add more solvent as necessary, and change collection containers. This automated technology has undoubtedly further increased the volume of solvents used and the volume of solvent waste generated in chromatography. A paper describing general methods for flash chromatography using disposable columns suggests an elution program based on 11 column volumes of solvent [11]. As an example for a 24 g column, which might be appropriate for purification of a 1.0 g sample, 11 column volumes corresponds to about 400 mL of solvent. This volume will generally dwarf the quantities of reactants and solvents used and waste generated from other aspects of a synthetic reaction step.

Several years ago and spurred by the availability of a commercial/ automated solvent distillation/recycling unit, this lab became interested in the recycling of solvents, especially the large volumes used in automated silica gel chromatography [12]. As binary ethyl acetate-hexane mixtures were generally being used as eluent, possibilities of recycling of these waste solvent mixtures were considered. As shown in table 1, the boiling points of ethyl acetate and hexane differ by only 8°C and they form an azeotrope containing 38% ethyl acetate [13]. Thus even if distillation could be efficient enough to achieve separation based on the small difference in boiling points, the azeotrope would not permit sufficient separation to be useful. Certainly, the mixture could simply be distilled and collected as a mixture, with additional hexane or ethyl acetate added to achieve the desired polarity. While this could be a viable strategy and it seems has gained some use, this is less attractive with automated chromatography systems that are designed to achieve the desired solvent ratio and to run gradients from pure solvents. Separation of ethyl acetate-hexane mixtures by azeotropic distillation with an added entrainer has been reported [14]. One study employed acetone as the entrainer, based on distillation of the lower boiling acetone-hexane azeotrope from the mixture [15]. This requires separation based on differences in boiling points of the acetone-hexane vs. ethyl acetate-hexane azeotropes of about 15°C (Table 1) and recovers the hexane as a mixture with acetone. A process based on enzymatic hydrolysis of the ethyl acetate component and recovery of the alkane component has also been described [16]. This approach would appear at best to be practical only for hexane recovery from mixtures containing only a small percentage of ethyl acetate or other ester solvent.

Table 1. Solvent boiling points and azeotropes[a]

Solvent A (b.p.)	Solvent B (b.p.)	b.p. (azeotrope)	% B in azeotrope
Hexane (69)	Ethyl acetate (77)	65	38
Heptane (98)	Ethyl acetate (77)		86.5
Hexane (69)	Acetone (56)	50	59
Heptane (98)	Acetone (56)	55.8	89.5
Octane (126)	Acetone (56)	no azeotrope	

[a] Adapted from ref. [12].

Table. 2. Trial distillations of solvent mixtures[a]

Solvent Mixture	Fraction	volume	% low b.p. component
ethyl acetate/heptane	<88°C	8.5 L	60%
ethyl acetate/heptane	>88°C	1.5 L	10%
acetone/heptane	<77 °C	7.0 L	69%
acetone/heptane	>77 °C	3.0 L	2.7%

[a] Both experiments used 10 L of a 1:1 mixture (by volume) of the two solvents.

Given the clear non-feasibility of separating ethyl acetate and hexane by distillation, a trial experiment was conducted for separating a mixture of ethyl acetate and heptane (Table 2). 5L each of ethyl acetate and heptane were combined and a distillation of the mixture was set up on the distillation unit. The instrument was programmed for changing collection containers at 88°C. The volume of the low boiling fraction collected was 8.5 L while the high boiling fraction was 1.5 L. Both fractions were analyzed by ^{1}H-NMR to determine the ratio of the two components in each. Even with such a low volume of the high boiling fraction, it still contained 10% ethyl acetate. This was not deemed to be a sufficient separation for the purpose of recycling. The experiment was repeated with 5L each of acetone and heptane. In this case, the collection was switched at a temperature of 77°C. The separation was significantly better, with 7L and 3L in the low boiling and high boiling fractions respectively. While the low boiling acetone fraction contained 31% heptane, the high boiling heptane fraction contained only 2.7% acetone. This was deemed acceptable for our purpose. The significant amount of heptane in the acetone fraction does limit the maximum eluting power that can be reached with recycled solvent. However, acetone is reported to have a somewhat higher eluting power relative to the commonly used ethyl acetate, which should at least partially compensate for the dilution effect [17]. Importantly the low concentration of acetone in the heptane fraction allows for a very low eluting power at the beginning of a solvent gradient, which is important when purifying very low-polarity compounds.

In assessing why acetone has not gained more use as a chromatographic solvent, a major issue has likely been the often stated UV cutoff of 300 nm. This value would appear valid in choosing a solvent for acquisition of UV spectra, and indeed acetone has a very strong absorbance at 254 nm, the common wavelength for detection in elution with ethyl acetate-alkane mixtures. In one report that evaluated more environmentally friendly solvents for chromatography, only solvents that do not absorb at 254 nm were considered, and thus acetone was excluded from the study [2]. However, acetone actually has a window of very low absorbance from about 205 to 220 nm. This allows detection at a much shorter wavelength than is possible with ethyl acetate, which permits detection of much simpler chromophores such as isolated carbon-carbon double bonds and ester groups. The UV absorbance profile of acetone is thus actually an advantage rather than a disadvantage in detecting UV active compounds from chromatography. This alone would be a reason for using acetone rather than ethyl acetate in many cases. The only foreseeable disadvantage of acetone in this regard would appear to be if the available instrument setup relies on a 254 nm fixed wavelength detector.

Flash chromatography in this lab was performed using a commercial automated chromatography system. The two detection wavelengths were chosen as 210 nm and 310 nm, with 210 nm being the primary wavelength which was used to direct fraction collection. For all compounds that exhibit a UV absorbance, eluent was sent to waste except for fractions exhibiting absorbance at 210 nm which were collected in 25 mL test tubes. The eluent sent directly to waste and fractions deemed to not contain the desired product were collected in a temporary waste container for recycling.

Solvent collected from rotary evaporation of product fractions were also transferred to the same waste container.

Solvent separation was achieved using a commercial automated recycling system. This system is specified as achieving 10 theoretical plates and can process a 19L batch in about 3 hours. For acetone-heptane mixtures, the lower-boiling acetone fraction was collected from 55 to 77°C and the higher boiling heptane fraction was collected from 77 to 120°C. The collected distillate was analyzed by H-NMR to assess purity and the acetone to heptane ratio.

Some practical considerations have been realized over the course of time of solvent recycling. Samples should be processed promptly after chromatography, combining and evaporating the desired fractions and transferring other fractions to waste collection. Tubes lose significant volume by evaporation if left sitting open for a day. This results in reduced recovery of solvent, in addition to environmental contamination. Simply covering the tops of the tubes with a flat object has limited effectiveness in reducing evaporation. Samples could be stored in a flammable storage refrigerator to reduce evaporation, if available. Small amounts of ethyl acetate or dichloromethane introduced in loading samples onto the column remain in the recycled solvent, but do not pose any problems in such small quantities. However, introduction of benzene, toluene, or other volatile aromatics into the waste stream is very problematic, as even small quantities of these substances in the elution solvent interfere with UV peak detection. The substitution of acetone for ethyl acetate in chromatography would appear to be fairly neutral in terms of environmental impact. Both acetone and ethyl acetate are on the preferred list (the best of three groupings) on the Pfizer solvent selection guide, [9c] are both in the "some issues" category in the GSK guide (the middle of three groupings), [9d] and have similar ratings on the ACS-GCI guide [9e] The primary environmental concerns regarding acetone stem from its volatility, which increases the likelihood of environmental release. However, use of a relatively low boiling solvent would appear to be essential for separation of a solvent mixture from chromatography by distillation. Use of a higher boiling polar solvent such as ethyl acetate would require an even higher boiling nonpolar component such as octane. Such higher boiling alkanes are more expensive, which would increase solvent cost. Perhaps more importantly, the high boiling point of higher alkanes would make the evaporation of solvent from product-containing fractions more difficult. The substitution of heptane for hexane is a clear improvement in terms of environmental and health and safety considerations. In the Pfizer solvent guide, heptane is on the usable list, while hexane and pentane are both on the undesirable list [9c] The GSK guide puts heptane in the "some issues" category while hexane is in the "major issues" category, [9d] and the ACS-GCI guide gives significantly better scores to heptane than to hexane [9e]. There seems to be substantial movement to use of heptane over hexane based on environmental and safety considerations, and the greater potential for recycling provides yet another reason to make this switch.

While acetone has not generally been a common solvent used in chromatography, it has long and widespread use in chemical and biological labs is in the cleaning of glassware. Acetone is effective for dissolving and removing residual organic substances, while its miscibility with water also lends itself to removal of residual water following standard sink-washing. The volatility of acetone also insures its rapid evaporation from the surface of glassware to avoid contamination by acetone in subsequent experiments. We also undertook the recycling of recovered acetone from glassware washing. Using the same commercial recycling instrument, the fraction boiling from 50 to 80°C was collected. The collected distillate contained small amounts of other volatile organic solvents and water, but was suitable for cleaning and final rinsing of glassware. The wash acetone waste was always collected and recycled separately from the chromatography waste, for two reasons. First the water in the acetone waste would be expected to distill with the heptanes, perhaps necessitating a drying step. Furthermore, the significant amount of heptane in the acetone fraction from the chromatography solvent makes it poorly suited for cleaning of glassware. There is a substantial startup cost to solvent recycling, with the recycling unit costing near $30,000. This cost might be recovered in a year in reduced solvent purchase in a setting where a batch of solvent is being processed on a daily basis, without even considering the waste disposal costs. Operational costs are minimal, the 10 amp/120 V rating of the instrument translating to less than $1.00 in electrical cost for one 19L process. The instrument uses air cooling, and thus does not require any water connection. There is some labor involved in the process, but it is not clear if this is greater than the labor involved in continually bringing in new solvent and more importantly, properly collecting and delivering solvent waste to a pickup location. Simpler much less expensive distillation/recycling setups may certainly be assembled and put into similar use [18]. A recent paper describes an automated solvent distillation unit home-built from readily available glassware and electronic components [19]. Built with a 10 round bottom boiling flask, the unit is convenient for recycling in 6L batches. Such a setup may be a good compromise in some cases between the substantial expense of the commercial unit and the amount of user attention and intervention required for a very basic distillation setup. Environmental and sustainability concerns dictate that solvent recycling is likely to become a requirement in any setting where substantial volumes of solvent are used. Acetone has the potential to play a major role in efforts to increase the degree of solvent recycling. In combination with heptane, acetone is an effective solvent for silica-based chromatography that allows for simple and efficient recycling. The low-wavelength spectral window of acetone may provide an advantage in UV detection relative to the more commonly used ethyl acetate. The traditional use of acetone as a solvent for cleaning of glassware is also readily coupled with recycling. In these and perhaps other applications, acetone should be a solvent of ever increasing importance for a sustainable future.

References

[1] Raymond, M. J.; Slater, C. S.; Savelski, M. J., LCA approach to the analysis of solvent waste issues in the pharmaceutical industry. *Green Chem* 2010, *12* (10), 1826-1834.

[2] Taygerly, J. P.; Miller, L. M.; Yee, A.; Peterson, E. A., A convenient guide to help select replacement solvents for dichloromethane in chromatography. *Green Chem* 2012, *14* (11), 3020-3025.

[3] Still, W. C.; Kahn, M.; Mitra, A., Rapid Chromatographic Technique for Preparative Separations with Moderate Resolution. *Journal of Organic Chemistry* 1978, *43* (14), 2923-2925.

[4] Katti, A. M.; Jagland, P., Development and optimization of industrial scale chromatography for use in manufacturing. *Analusis* 1998, *26* (7), M38-M46.

[5] Jageland, P.; Magnusson, J.; Bryntesson, M., Optimization of Industrial-Scale High-Performance Liquid-Chromatography Applications Using a Newly Developed Software. *J Chromatogr A* 1994, *658* (2), 497-504.

[6] Welch, C. J.; Leonard, W. R.; DaSilva, J. O.; Biba, M.; Albaneze-Walker, J.; Henderson, D. W.; Laing, B.; Mathre, D. J., Preparative chiral SFC as a green technology for rapid access to enantiopurity in pharmaceutical process research. *Lc Gc Eur* 2005, *18* (5), 264-+.

[7] Welton, T., Solvents and sustainable chemistry. *P Roy Soc a-Math Phy* 2015, *471* (2183).

[8] Peterson, E. A.; Dillon, B.; Raheem, I.; Richardson, P.; Richter, D.; Schmidt, R.; Sneddon, H. F., Sustainable chromatography (an oxymoron?). *Green Chem* 2014, *16* (9), 4060-4075.

[9] (a) Prat, D.; Pardigon, O.; Flemming, H. W.; Letestu, S.; Ducandas, V.; Isnard, P.; Guntrum, E.; Senac, T.; Ruisseau, S.; Cruciani, P.; Hosek, P., Sanofi's Solvent Selection Guide: A Step Toward More Sustainable Processes. *Org Process Res Dev* 2013, *17* (12), 1517-1525; (b) Capello, C.; Fischer, U.; Hungerbuhler, K., What is a green solvent? A comprehensive framework for the environmental assessment of solvents. *Green Chem* 2007, *9* (9), 927-934; (c) Alfonsi, K.; Colberg, J.; Dunn, P. J.; Fevig, T.; Jennings, S.; Johnson, T. A.; Kleine, H. P.; Knight, C.; Nagy, M. A.; Perry, D. A.; Stefaniak, M., Green chemistry tools to influence a medicinal chemistry and research chemistry based organisation. *Green Chem* 2008, *10* (1), 31-36; (d) Henderson, R. K.; Jimenez-Gonzalez, C.;

Constable, D. J. C.; Alston, S. R.; Inglis, G. G. A.; Fisher, G.; Sherwood, J.; Binks, S. P.; Curzons, A. D., Expanding GSK's solvent selection guide - embedding sustainability into solvent selection starting at medicinal chemistry. *Green Chem* 2011, *13* (4), 854-862; (e) *ACS GCI Pharmaceutical Roundtable Solvent Selection Guide*. https://www.acs.org/content/acs/en/greenchemistry/research-innovation/tools-for-green-chemistry.html (accessed Nov. 20, 2017).

[10] Brown, L.; Earle, M. J.; Gilea, M. A.; Plechkova, N. V.; Seddon, K. R., Ionic Liquid-Liquid Chromatography: A New General Purpose Separation Methodology. *Topics Curr Chem* 2017, *375* (5).

[11] Stevens, W. C.; Hill, D. C., General methods for flash chromatography using disposable columns. *Mol Divers* 2009, *13* (2), 247-252.

[12] Drueckhammer, D. G.; Gao, S. Q.; Liang, X. F.; Liao, J. Z., Acetone-Heptane as a Solvent System for Combining Chromatography on Silica Gel with Solvent Recycling. *ACS Sustainable Chemistry & Engineering* 2013, *1* (1), 87-90.

[13] Horsley, L. H., *Azeotropic Data - III*. American Chemical Society: Washington, 1973; Vol. 116.

[14] Rodriguez-Donis, I.; Acosta-Esquijarosa, J.; Gerbaud, V.; Pardillo-Fondevila, E.; Joulia, X., Separation of n-hexane-ethyl acetate mixtures by azeotropic batch distillation with heterogeneous entrainers. *Chem Eng Process* 2005, *44* (1), 131-137.

[15] Acosta, J.; Arce, A.; Martinez-Ageitos, J.; Rodil, E.; Soto, A., Vapor-liquid equilibrium of the ternary system ethyl acetate plus hexane plus acetone at 101.32 kPa. *J Chem Eng Data* 2002, *47* (4), 849-854.

[16] Andrade, L. H.; Vitta, P. B. D.; Utsunomiya, R. S.; Schoenlein-Crusius, I. H.; Porto, A. L. M.; Comasseto, J. V., Bio-recycling of hexanes from laboratory waste mixtures of hexanes and ethyl acetate by ester biotransformation: a green alternative process. *Appl Catal B-Environ* 2005, *59* (3-4), 197-203.

[17] Sadek, P. C., *The HPLC Solvent Guide*. Wiley-Interscience: New York, 2002; (b) Snyder, L. R., *Mobile-Phase Effects in Liquid-Solid Chromatography*. Academic Press: New York, 1983; Vol. 3, p 157-

223; (c) Marini-Bettolo, G. B., *Thin-Layer Chromatography of Steroids*. Elsevier: New York, 1964.

[18] Weires, N. A.; Johnston, A.; Warner, D. L.; McCormick, M. M.; Hammond, K.; McDougal, O. M., Recycling of Waste Acetone by Fractional Distillation. *J Chem Educ* 2011, *88* (12), 1724-1726.

[19] Zweckmair, T.; Hell, S.; Klinger, K. M.; Rosenau, T.; Potthast, A.; Bohmdorfer, S., Recycling of Analytical Grade Solvents on a Lab Scale with a Purpose-Built Temperature-Controlled Distillation Unit. *Org Process Res Dev* 2017, *21* (4), 578-584.

INDEX

A

B

E

F

J

K

L

M

N

O

S

T

U

V

W

X

Z